THE PHYSICS OF MASSIVE NEUTRINOS

World Scientific Lecture Notes in Physics Vol. 25

THE PHYSICS OF MASSIVE NEUTRINOS

BORIS KAYSER
National Science Foundation

with

FRANÇOISE GIBRAT-DEBU
CEN-Saclay

and

FREDERIC PERRIER
Stanford Linear Accelerator Center

World Scientific
Singapore • New Jersey • London • Hong Kong

Published by

World Scientific Publishing Co. Pte. Ltd.
P O Box 128, Farrer Road, Singapore 9128

USA office: World Scientific Publishing Co., Inc.
687 Hartwell Street, Teaneck, NJ 07666, USA

UK office: World Scientific Publishing Co. Pte. Ltd.
73 Lynton Mead, Totteridge, London N20 8DH, England

Library of Congress Cataloging-in-Publication data is available.

THE PHYSICS OF MASSIVE NEUTRINOS

ISBN 9971-50-661-0
9971-50-662-9 pbk

Printed in Singapore by JBW Printers & Binders Pte. Ltd.

PREFACE

These lectures, given originally at Saclay in November 1985, are intended to explain the physics and phenomenology of massive neutrinos. They require of the reader only a knowledge of quantum mechanics and of very elementary quantum field theory.

After arguing that neutrino mass is not unlikely, and briefly considering the search for evidence of this mass in decay processes, we examine the physics and phenomenology of neutrino oscillation. Then we discuss the physics of Majorana neutrinos (neutrinos which are their own antiparticles). We first establish a number of basic properties of such neutrinos without using any field theory. We then introduce the field-theory description of a Majorana particle, and use it to treat such processes as neutrinoless double beta decay. Finally, having studied Majorana neutrinos, we turn to the treatment of neutrino masses in gauge theories, and derive the "see-saw relation", which explains why neutrino masses are so small by relating them to the inverse of a large mass scale.

Our references to experimental results and to the literature were compiled in March 1987, and updated in July 1988.

Boris Kayser
Frédéric Perrier
Françoise Gibrat-Debu

TABLE OF CONTENTS

THE PHYSICS OF MASSIVE NEUTRINOS

CHAPTER

ONE

INTRODUCTION

Why should we talk about neutrino masses? There is no incontrovertible experimental evidence that neutrinos have mass. Nevertheless, some physicists believe that they are in fact massive.

A good reason for this belief is that from the standpoint of the grand unified theories (GUTS), which try to unify the description of the weak, electromagnetic, and strong interactions, it is more natural for neutrinos to be massive than to be massless. In any grand unified theory, we put a given neutrino ν in a large multiplet F together with at least one charged lepton ℓ, one positively-charged quark q^+, and one negatively-charged quark q^-:

$$F = \begin{bmatrix} q^+ \\ q^- \\ \ell \\ \nu \end{bmatrix}$$

In this scheme, neutrinos become brothers of leptons and quarks which are all massive. Thus, it would be exceptional for them to be massless! Of course, this is not a proof that neutrinos have mass. However, it does make it very natural for them to do so.

Here we shall assume that neutrinos do indeed have mass. To be sure, there is as yet no very good evidence for the grand unified theories.

Proton decay, which is perhaps the most spectacular prediction of these models, has not been observed, and the proton lifetime predicted by the minimal SU(5) GUT is ruled out by experiment. However, this particular GUT is a somewhat unnatural one, since it does not place all the quarks and leptons of one generation in a single multiplet, but in a package consisting of two multiplets. Grand unified theories as a class are very much alive. In fact, the famous and popular Superstring Theories strongly suggest that, at the appropriate energy scale, Nature is indeed described by a grand unified theory.

Let us return to the multiplet F. We know experimentally that the neutrino mass is much smaller than the quark and lepton masses:

$$M_\nu \ll M_{q^-}, M_{q^+}, M_\ell$$

How does the neutrino get to be much lighter than all the other fermions? Perhaps, unlike q^+, q^-, ℓ, which carry electric charge, the neutrino is its own antiparticle, and perhaps this circumstance is the origin of its relative lightness.

But what about the lepton number? Does not this quantum number distinguish a $\bar{\nu}$ from a ν? To see why it may not do that, let us recall the origin of the introduction of this number. It is observed experimentally that the neutral particle which is emitted simultaneously with a μ^+ in the π^+ decay interacts with matter producing only a μ^- and never a μ^+. This neutral particle is called ν_μ.

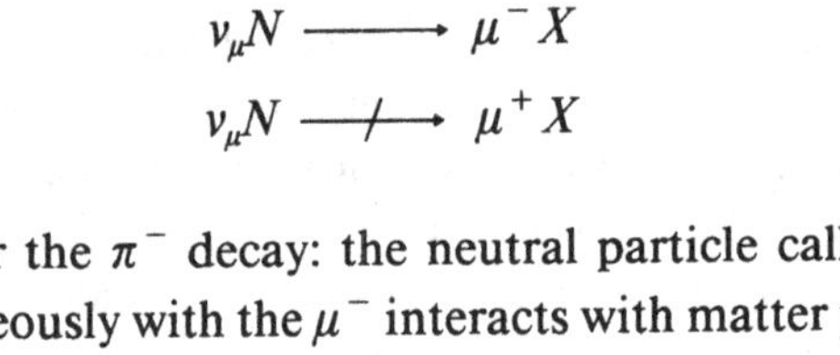

$$\nu_\mu N \longrightarrow \mu^- X$$
$$\nu_\mu N \not\longrightarrow \mu^+ X$$

It is similar for the π^- decay: the neutral particle called $\bar{\nu}_\mu$ which is emitted simultaneously with the μ^- interacts with matter producing only a μ^+ and never a μ^-:

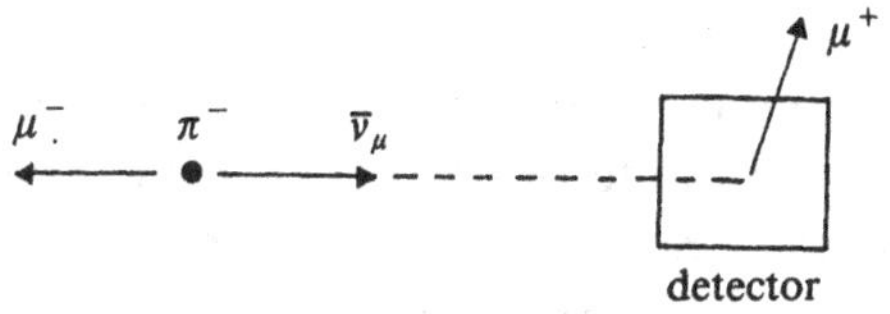

$$\bar{\nu}_\mu N \longrightarrow \mu^+ X$$

$$\bar{\nu}_\mu N \not\longrightarrow \mu^- X$$

The conventional explanation for these facts is the following:

(i) ν_μ and $\bar{\nu}_\mu$ are distinct from each other.

(ii) There is a quantum number which is conserved during these interactions: the lepton number. We assign the lepton number $+1$ to μ^- and ν_μ and -1 to μ^+ and $\bar{\nu}_\mu$: then ν_μ can only produce μ^- in matter.

However, there is no evidence that this explanation is the correct one. Indeed, the existence of the lepton number is not necessary.

An alternative explanation arises from the fact that the neutral particle produced in the π^+ decay has a left-handed helicity, while the neutral particle produced in the π^- decay has a right-handed helicity. Now, suppose that the parity-violating weak interaction is such that every neutrino with a right-handed helicity interacts giving a μ^+, while every neutrino with a left-handed helicity interacts giving a μ^-. In this case we don't need to introduce a conserved lepton number! The helicity constraints are sufficient to account for all the observed facts. If they are the true explanation, then, in contradiction with (i), "ν_μ" and "$\bar{\nu}_\mu$" are, respectively, just the left and right helicity states of a single particle ν^M (M for Majorana). In contrast, if the hypothesis (i)–(ii) is the true explanation, then ν_μ is a so-called Dirac neutrino ν^D. This is a particle with two helicity states whose antiparticle $\bar{\nu}^D$ (here $\bar{\nu}_\mu$) differs from ν^D and has two helicity states of its own, for a total of four states. We will explain the difference between Dirac and Majorana neutrinos in much greater detail later.

To tell whether "ν_μ" and "$\bar{\nu}_\mu$" are just the two helicity states of a Majorana neutrino, or do truly differ in some way other than helicity, one would like to carry out the following experiment: Produce left-handed ν_μ particles in π^+ decays. Somehow reverse the helicities of these particles, turning them into right-handed objects. Now find out whether these right-handed objects interact with matter in the same way as the right-handed "$\bar{\nu}_\mu$" particles produced in π^- decays (Majorana case), or not (Dirac case). Unfortunately, this experiment would be extremely difficult, and it has not been done.

In the most popular explanation of the lightness of neutrinos relative to the charged leptons and quarks, a particle which starts out as a Dirac neutrino (four states: ν^D and $\bar{\nu}^D$, each with two helicities) splits into two Majorana neutrinos (two states each). One of these Majorana neutrinos, ν, is light, and is identified as the familiar neutrino, while the other one, N, is heavy. In this scheme, which is due to Gell-Mann-Ramond-Slansky, Yanagida, and Mohapatra-Senjanovic,[1] it is natural for the ν and N masses to be related by

$$M_\nu M_N = M^2_{q \text{ or } \ell} \tag{1.1}$$

where $M_{q \text{ or } \ell}$ is a typical quark or charged lepton mass. Furthermore, M_N is related to a symmetry-breaking scale, and so is in general quite big. Thus we can conclude that $M_\nu \lll M_{q \text{ or } \ell}$.

Using relation (1.1), we can get some orders of magnitude: if $M_N \sim M_{\text{GUT}} = 10^{14}$ GeV and $M_{q \text{ or } \ell} \sim 1$ GeV then $M_\nu \sim 10^{-5}$eV! We can have neutrino masses much smaller than what can now be measured! If $M_N \sim 100$ GeV and $M_{q \text{ or } \ell} \sim 1$ MeV we obtain a mass M_ν of the order of magnitude of the experimental limits.

Recall that the mass of the neutrino has fundamental consequences in astrophysics: if the mass of the neutrino is of order of 10 eV and if its lifetime is large compared to the age of the universe, the neutrinos could dominate the mass density of the universe.

We would like to end this introduction by emphasizing the importance of trying to find new experiments that would be sensitive to neutrino masses well below 1 eV!

CHAPTER

TWO

PHENOMENOLOGY OF MASSIVE NEUTRINOS

Now, we would like to go through some phenomenological consequences of neutrino mass, focussing on how one can look for evidence of it. To make things clear, let us define the following simple theoretical framework.

1 - A SIMPLE THEORETICAL FRAMEWORK

Assume that we have three charged leptons, e, μ and τ. Each of these leptons ℓ is coupled to a neutral state called ν_ℓ (neutrino associated with the lepton ℓ) via the intermediate vector boson discovered at CERN with a mass around 83 GeV. The Hamiltonian density for the charged current weak interaction is then (g is a coupling constant):

$$H_{CC} = gW_\mu(83\ \mathrm{GeV}) \sum_{\ell=e,\mu,\tau} i\bar{\ell}\gamma_\mu(1+\gamma_5)\nu_\ell + h.c. \qquad (2.1)$$

The set $\{\nu_\ell\}$ is said to be the set of flavor eigenstates or weak interaction eigenstates.

Now assume that the freely propagating neutrinos of definite mass are not the ν_ℓ but a set of three neutral states ν_m with masses M_m, and that the flavor eigenstates ν_ℓ are linear combinations of these mass eigenstates ν_m:

$$\nu_\ell = \sum_m U_{\ell m}\nu_m \qquad (2.2)$$

Here U is the neutrino mixing matrix. We have

$$H_{CC} = gW_\mu(83\text{ GeV}) \sum_{\ell=e,\mu,\tau} i\bar{\ell}\gamma_\mu(1+\gamma_5)\sum_m U_{\ell m}\nu_m + h.c.$$

The mixing matrix U is the analog of the Kobayashi-Maskawa matrix in the charged current Hamiltonian between quarks:

$$H_{CC} = gW_\mu(83\text{ GeV})\sum_i i\bar{q}_i^{\,+}\gamma_\mu(1+\gamma_5)\sum_j (KM)_{ij}q_j^- + h.c. \quad (2.3)$$

where: i and j are generation indices
q^+ is a positive quark u, c, or t
q^- is a negative quark d, s, or b
KM is the Kobayashi-Maskawa matrix

The existence of the quark mixing, described by the Kobayashi-Maskawa matrix, has been established experimentally. It is then natural to expect that there is lepton mixing, described by the matrix U, as well. The mixing matrix U has the following properties:

(i) In a gauge theory, it is a unitary matrix: it relates the basis $\{\nu_m\}$ of the eigenstates of the free-particle Hamiltonian to the basis $\{\nu_\ell\}$ of the eigenstates of the gauge group.

(ii) If CP is conserved in our theory, the matrix U can be chosen real. We will explain this later on.

Let us now describe simple phenomena involving massive neutrinos and firstly: decays.

2 - WHAT WE HAVE LEARNT FROM DECAYS

Consider a decay involving a neutrino in the final state:

$$X \rightarrow Y + \nu$$

Due to the neutrino mixing, this decay is actually the incoherent sum of decays into all the neutrino mass eigenstates whose emission is kinematically possible:

$$X \to Y + \nu = \sum_m (X \to Y + \nu_m)$$

For example, the decay $\pi \to \mu\nu_\mu$ is actually:

$$\pi \to \mu + \nu_\mu = \sum_m (\pi \to \mu + \nu_m)$$

The spectrum of the final muon momentum will then have several monochromatic lines, each of these corresponding to a mass eigenstate ν_m, and having a strength proportional to $|U_{\mu m}|^2$ (Ref. 2). People have looked at this spectrum, and seen no lines. Taking into account the experimental resolution, they have obtained limits on $|U_{\mu m}|^2$ for ν_m masses above ~ 10 MeV.[3] In addition, study of the decay $\pi \to \mu + \nu_\mu$ has shown that the mass of the eigenstate ν_m which is emitted most of the time in the decay is less than 250 keV.[4] This statement is shortened to:

$$M(\nu_\mu) < 250 \text{ keV}$$

Information on the mass eigenstate which is the dominant piece of the flavor eigenstate ν_e is coming from experiments on tritium β decay:

$$\mathrm{T} \to {}^3\mathrm{He} + e^- + \bar{\nu}_e$$

An experiment in Moscow claims that the dominant mass eigenstate in this decay has a mass between 17 and 40 eV.[5] But this result is quite controversial. A number of other experiments on tritium β decay are in progress.[6,7]

Limits on the τ neutrino mass, the mass of the eigenstate ν_m mostly coupled to the τ lepton, have been obtained from the decay of τ leptons produced in e^+e^- collisions. The most stringent limit comes from study of the 5π invariant mass distribution in the reaction:

$$e^+e^- \to \tau^+\tau^-, \quad \tau^- \to (5\pi)^- \nu_\tau$$

The result is (Ref. 8):

$$M(\nu_\tau) < 35 \text{ MeV}$$

In all the processes above, the distinction between Majorana and Dirac neutrinos is irrelevant.

A very interesting process is the rare decay of the kaon into one pion and two neutrinos (this decay is not yet observed):[9]

$$K^+ \to \pi^+ \nu\bar{\nu} = \sum_m (K^+ \to \pi^+ \nu_m \bar{\nu}_m)$$

This decay has the very special property that it is, in principle, sensitive to the Dirac or Majorana character of the neutrinos.[10]

The only thing that can be measured here is the energy of the outgoing pion. One can show that, due to the unitarity of the U matrix, the pion energy spectrum does not depend on the $U_{\ell m}$. However, as shown in Figure 1,[10] this spectrum *does* depend on the neutrino mass (we assume only one ν_m has a mass that is significant). In addition to this, if the emitted neutrinos are nonrelativistic (if the mass of the heaviest mass eigenstate is large enough), the distortion of the spectrum is very different if these neutrinos are Dirac or Majorana particles (see Figure 1(a)). If, however, the neutrinos are relativistic (neutrino masses very small compared to a typical momentum), it becomes very hard to tell whether they are Majorana or Dirac particles (see Figure 1(b)).

This is, in fact, an example of a very general phenomenon which we refer to as the Majorana-Dirac confusion theorem:[11] Assuming there are no right-handed currents, if the neutrino masses are very small compared to the experimental energy scale, it is impossible to tell the difference between a Dirac neutrino and a Majorana neutrino. We shall come back to this theorem later.

Besides decays, a very important consequence of the neutrino masses is the possible existence of neutrino oscillations.

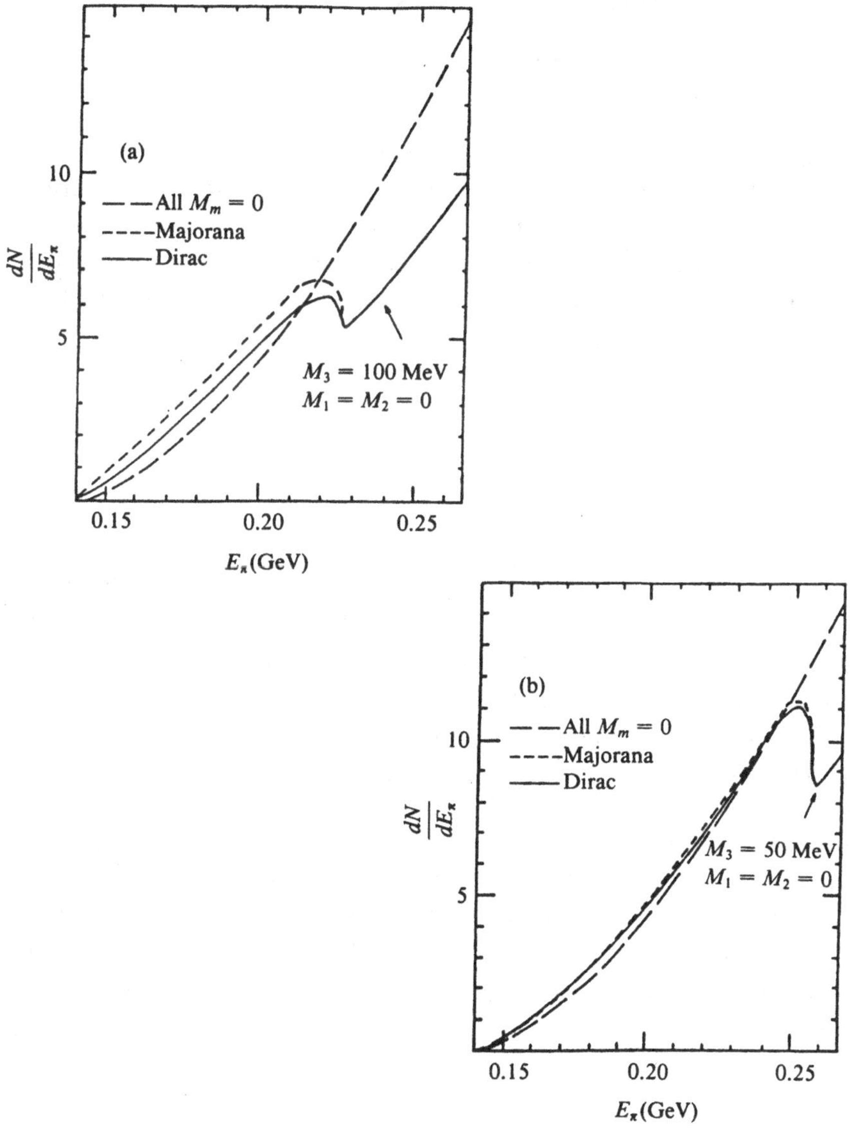

Fig. 1. Effect of the neutrino mass on the decay $K \rightarrow \pi\nu\bar{\nu}$.

3 - THE NEUTRINO OSCILLATIONS

a) Neutrino oscillations in vacuum

i) Principle of neutrino oscillations

Let us describe a typical neutrino-oscillation experiment. Take a pion beam and let the pions decay into muons and neutrinos (by definition, see chapter II.1, these neutrinos are called muon neutrinos ν_μ). Downstream of the decaying pions is a target-detector looking for neutrino interactions. Suppose that we see an interaction with an *electron* in the final state:

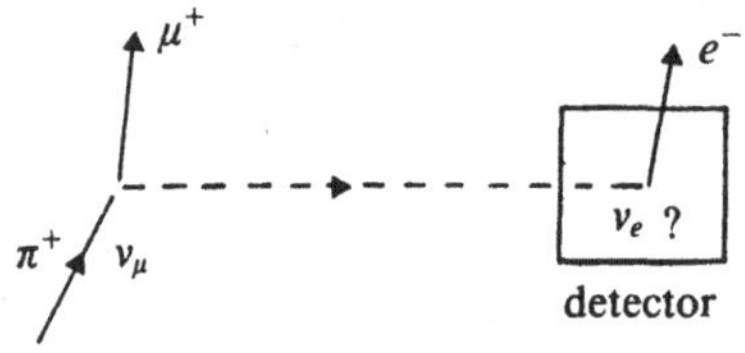

The trouble is that the neutrino which is able to produce an electron is not ν_μ but ν_e. The only explanation is that our muon neutrino ν_μ has transformed itself into an electron neutrino ν_e between the decay point and the detector. How is that possible?

The answer is given by quantum mechanics. Assume, to be quite general, that we have N flavors of charged leptons $\ell = e, \mu, \tau, \ldots$ associated with N neutrinos $\nu_e, \nu_\mu, \nu_\tau, \ldots$ Suppose that we have a source of ν_ℓ neutrinos. We said that the flavor eigenstate ν_ℓ is not a mass eigenstate but a linear superposition of mass eigenstates ν_m. Assume that there are N ν_m, and that the mass resolution of the experiment is much poorer than the level of sensivity, $M_m - M_{m'}$, needed to distinguish one ν_m from another. Then each neutrino produced by our source is the coherent superposition:

$$\nu_\ell = \sum_{m=1}^{N} U_{\ell m}\, \nu_m$$

of the mass eigenstates.

Let us assume for the sake of simplicity that we have a neutrino ν_ℓ born at time $t = 0$ with momentum p_ν perfectly defined. At this time, the wave function is therefore:

$$\psi(x, t = 0) = \sum_m U_{\ell m} \, \nu_m \, e^{ip_\nu x} \tag{2.4}$$

After a time t, this will evolve into:

$$\psi(x, t) = \sum_m U_{\ell m} \, \nu_m \, e^{ip_\nu x} \, e^{-iE_m t},$$

with
$$E_m = E(\nu_m) = \sqrt{p_\nu^2 + M_m^2}$$

Assuming $M_m \ll p_\nu$,
$$E_m \simeq p_\nu + \frac{M_m^2}{2p_\nu}$$

With all $M_m \ll p_\nu$, our ν is travelling almost with the speed of light. Hence, if it was born at $x = 0$, then at time t, it will be approximately at $x = t$. Let us then examine $\psi(t, t) = \psi(x, x)$, which is given by:

$$\psi(x, x) \simeq \sum_m U_{\ell m} \, \nu_m \, e^{-i[M_m^2/2p_\nu]x}$$

Expressing ν_m as a combination of the ν_ℓ's, we find:

$$\nu_m = \sum_{\ell'} U^*_{\ell' m} \, \nu_{\ell'}$$

so:

$$\psi(x, x) = \sum_{\ell'} \left[\sum_m U_{\ell m} \, e^{-i(M_m^2/2p_\nu)x} \, U^*_{\ell' m} \right] \nu_{\ell'} \tag{2.5}$$

We see that this wave function is a superposition of all the neutrino flavors. In it, the amplitude for our neutrino, born with flavor ℓ, to have a new flavor ℓ' after travelling a distance x is just the coefficient of $\nu_{\ell'}$.

Note that, as illustrated in Figure 2, this amplitude has the same structure as that for a multi-slit interference experiment. It consists of an amplitude for the original ν_ℓ to be a ν_m, times an amplitude for the ν_m, having propagated, to be a $\nu_{\ell'}$. This product of amplitudes is then summed coherently over all possible "slits" ν_m.

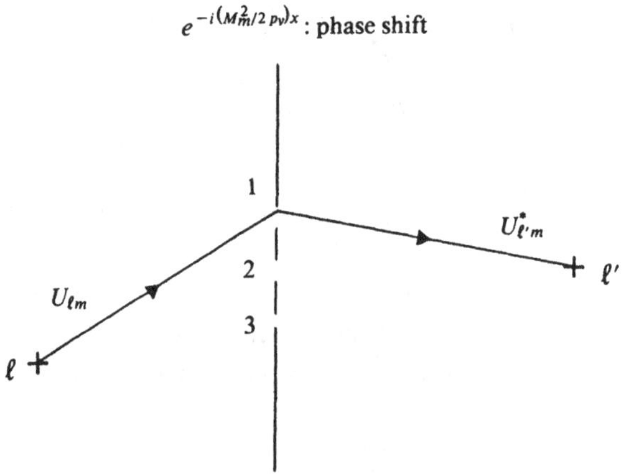

Fig. 2. Multi-slit interference experiment. The structure of the amplitude of the neutrino oscillation is very general in quantum mechanics.

Neutrino oscillation can be understood in a very simple way. What happens is that, at a given p_ν, the lighter mass states in the original ν_ℓ travel faster than the heavier ones, and get ahead of the latter. Thus, the various ν_m components of the beam get out of phase with one another, and do not add up to a ν_ℓ anymore. Thus, as it travels, the beam picks up components corresponding to other flavors.

Let $P(\ell \to \ell', x)$ be the probability of finding the neutrino to have flavor ℓ' at a distance x from its source, if originally it had flavor ℓ. From Eq. (2.5), we have:

$$P(\ell \rightarrow \ell', x) = \left[\sum_{m'} U^*_{\ell m'}\, e^{iM^2_{m'} x/2p_\nu}\, U_{\ell' m'}\right] \cdot \left[\sum_{m} U_{\ell m}\, e^{-iM^2_{m} x/2p_\nu}\, U^*_{\ell' m}\right]$$

$$= \sum_m |U_{\ell m}|^2 |U_{\ell' m}|^2 \tag{2.6}$$

$$+ \sum_{m' \neq m} Re(U_{\ell m} U^*_{\ell m'} U_{\ell' m'} U^*_{\ell' m}) \cos\left(\frac{M^2_m - M^2_{m'}}{2p_\nu} x\right)$$

$$+ \sum_{m' \neq m} Im(U_{\ell m} U^*_{\ell m'} U_{\ell' m'} U^*_{\ell' m}) \sin\left(\frac{M^2_m - M^2_{m'}}{2p_\nu} x\right)$$

Now, we make the assumption that *CP is conserved*: one can then *choose U real*, and the previous expression simplifies to:

$$P(\ell \rightarrow \ell', x) = \sum_m U^2_{\ell m} U^2_{\ell' m} + \sum_{m' \neq m} U_{\ell m} U_{\ell m'} U_{\ell' m'} U_{\ell' m} \cos\left(2\pi \frac{x}{L_{mm'}}\right) \tag{2.7}$$

This probability has a beautiful oscillatory pattern as a function of distance x. In it, the quantities $L_{mm'}$ are:

$$L_{mm'} = 2\pi \frac{2p_\nu}{|M^2_m - M^2_{m'}|} \equiv 2\pi \frac{2p_\nu}{\delta M^2_{mm'}} \tag{2.8}$$

$L_{mm'}$ is the *oscillation length* between ν_m and $\nu_{m'}$.

Note that:

– If all the masses are equal (in particular, if they all vanish) there is no oscillation.

– If $\nu_{\ell_o} = \nu_{m_o}$, ν_{ℓ_o} doesn't oscillate into $\nu_{\ell'}$ with $\ell' \neq \ell_o$.

That is, *oscillation requires neutrinos to have both mass and non-trivial mixing.*

Let us now make some comments on the features of the neutrino oscillations:

Remarks:

0 – The oscillatory terms in $P(\ell \to \ell', x)$ come from interference between the different mass eigenstates in the neutrino wave function.

1 – If $x \ll L_{mm'}$, ν stays in its original flavor.

2 – If x becomes too large ($x \gg L_{mm'}$), the oscillation pattern will be washed out. This is due to the spread in momentum Δp_ν in any actual neutrino beam. More precisely, the oscillations will disappear if the neutrino with momentum p_ν has a phase shift $\sim \pi$ compared to the neutrino with momentum $p'_\nu = p_\nu + \Delta p_\nu/2$. The average over the oscillatory term will then vanish.

Let X be the wash-out distance and $L'_{mm'}$ the oscillation length corresponding to the momentum p'_ν. We have:

$$2\pi \frac{X}{L'_{mm'}} = 2\pi \frac{X}{L_{mm'}} - \pi$$

$$L'_{mm'} \simeq L_{mm'}\left(1 + \frac{\Delta p_\nu}{2p_\nu}\right)$$

therefore:

$$X \sim \frac{p_\nu}{\Delta p_\nu} L_{mm'} \tag{2.9}$$

The oscillations disappear if the distance x is bigger than X. Beyond that point we have:

$$P(\ell \to \ell', x) = \sum_m U^2_{\ell m} U^2_{\ell' m} \neq 0$$

Notice that it is still possible to find an ℓ' neutrino in the ℓ neutrino beam, but the probability for this will no longer vary with distance.

3 – As a conclusion, the oscillation pattern can be observed if the length of the experiment x is of the order of magnitude of the oscillation length $L_{mm'}$.

There are basically two types of neutrino oscillation experiments:

- *Appearance experiments*: these look for $\nu_\ell \to \nu_{\ell'}$, where $\ell' \neq \ell$.

- *Disappearance experiments*: these look for a reduction in the flux of neutrinos of the original flavor, or else for an x-dependence of this flux.

ii) Neutrino oscillations and the principles of quantum mechanics

As already stressed, the phenomenon of oscillation between a set of observable states is a basic feature of quantum mechanics. It appears, for example, when a polarized light beam goes through a birefringent crystal or matter made out of chiral molecules. There is also a famous example in particle physics: the neutral kaon system.

We would like now to illustrate that neutrino oscillation is a lovely example of quantum mechanics in action.[12]

Take our typical neutrino oscillation experiment pictured in the figure at the beginning of the previous section. Suppose we see electrons produced in our detector by the neutrinos which started out as muon neutrinos. Suppose, further, that we see the probability $P(\nu_\mu \rightarrow \nu_e, x)$ oscillating with distance as predicted by formula (2.7).

Now, suppose that we can detect the muons from the pion decays and measure their momenta accurately enough that we know which mass eigenstate ν_m was actually emitted in each decay. This is, of course, impossible in practice (for the moment at least), but let us continue with this "Gedankenexperiment". Now, if we know which ν_m actually goes down the beam line in each event, the beam state is no longer a coherent superposition of mass eigenstates: a particular one has been selected. Therefore, the probability of producing a ν_e can no longer oscillate with distance, because such oscillation requires interference between different mass eigenstates. The measurements which tell us which ν_m is emitted in each pion decay have destroyed the oscillation pattern. It is interesting to understand how they cause this destruction.

The answer is given by the uncertainty principle:[12]

In order to determine the invariant mass M_ν^2 of the decay neutrino, one has to measure the momenta of both the pion and the muon. The uncertainty principle says that if the momentum of the pion is measured very precisely, the information about its position (the decay point) is lost. More quantitatively, if E_ν and p_ν are, respectively, the energy and momentum of the neutrino, so that $M_\nu^2 = E_\nu^2 - p_\nu^2$, then the error in the M_ν^2 measurement is given by:

$$\Delta(M_\nu^2) = [(2E_\nu)^2 (\Delta E_\nu)^2 + (2p_\nu)^2 (\Delta p_\nu)^2]^{1/2}$$

Here ΔE_ν and Δp_ν are the errors in the E_ν and p_ν measurements, and we are assuming them to be uncorrelated.

Now, it will be possible to determine which mass eigenstate is emitted in each decay only if $\Delta(M_\nu^2)$ is less than the smallest mass squared difference $|M_m^2 - M_{m'}^2|$, which requires that:

$$2p_\nu \Delta p_\nu < |M_m^2 - M_{m'}^2|$$

The uncertainty in the position of the source is then:

$$\Delta x > \frac{2p_\nu}{|M_m^2 - M_{m'}^2|} = \frac{1}{2\pi} L_{mm'}$$

We see that the uncertainty in the position of the source is then bigger than the largest oscillation length, and all oscillations are washed out. If you have identified which mass eigenstate you have produced, you don't know where the source is accurately enough to see oscillations!

If the M_m are of the order of 10 eV and p_ν is bigger than 100 MeV, one will certainly not *accidentally* measure p_ν accurately enough to wipe out the oscillation pattern. But one may want to look for neutrino masses of order 100 MeV in pion or kaon decay precisely by finding the mass of the ν_m emitted in each meson decay. The identified ν_m won't oscillate, of course. Furthermore, for $M_m = 100$ MeV, and $M_{m'} \simeq 0$, the oscillation length would be less than 10^{-7} cm, even if p_ν is as big as 1 TeV.

To see oscillation with distance, the source of the neutrino beam must be localized. Thus, p_ν cannot be sharp. A wave packet treatment should really be given to account for the spread in momentum. This has been done,[12] and it gives the same result as the simplified calculation we have done here. It also gives a justification of the assumption we made, namely that the different ν_m in a beam have a common momentum p_ν and different energies E_m, rather than, say, a common energy E_ν and different momenta p_m. More details on these problems can be found in the paper referred to above.

iii) Two-neutrino oscillation

Suppose that only two neutrino flavor eigenstates, ν_{ℓ_a} and ν_{ℓ_b}, and two mass eigenstates, ν_1 and ν_2, participate appreciably in the neutrino mixing. Then:

$$\begin{aligned} \nu_{\ell_a} &= \nu_1 \cos\theta + \nu_2 \sin\theta \\ \nu_{\ell_b} &= -\nu_1 \sin\theta + \nu_2 \cos\theta \end{aligned}$$

so that the mixing matrix U is $\begin{bmatrix} \cos\theta & \sin\theta \\ -\sin\theta & \cos\theta \end{bmatrix}$, a 2×2 rotation matrix.

There is now only one oscillation length, namely $L = \dfrac{4\pi p_\nu}{|M_2^2 - M_1^2|} = \dfrac{4\pi p_\nu}{\delta M^2}$.

From Eq. (2.7), the probability $P(\nu_{\ell_a} \to \nu_{\ell_b}, x)$ that would be studied in an appearance experiment is given by:

$$P(\nu_{\ell_a} \to \nu_{\ell_b}, x) = \sin^2 2\theta \sin^2 \pi \frac{x}{L} = \sin^2 2\theta \sin^2 \delta_{12} \tag{2.10}$$

$$\delta_{12} = \pi \frac{x}{L} = \frac{\delta M^2 x}{4 p_\nu}$$

By conservation of probability, the probability $P(\nu_{\ell_a} \to \nu_{\ell_a}, x)$ that would be studied in a disappearance experiment is then given by:

$$P(\nu_{\ell_a} \to \nu_{\ell_a}, x) = 1 - \sin^2 2\theta \sin^2 \pi \frac{x}{L} \tag{2.11}$$

We would like to use real units instead of the natural units where $\hbar = c = 1$. The apparent dimension of the dimensionless δ_{12} is [length]·[energy], so

$$\delta_{12} = \frac{1}{4}\frac{1}{\hbar c}\frac{\delta M^2 x}{p_\nu} = \frac{1}{4}\frac{10^{18}\cdot 10^{-18}}{0.197}\frac{\dfrac{\delta M^2}{1\text{eV}^2}\cdot\dfrac{x}{1\text{km}}}{\dfrac{p_\nu}{1\text{GeV}}}$$

or

$$\delta_{12} = 1.27\frac{\dfrac{\delta M^2}{1\text{eV}^2}\cdot\dfrac{x}{1\text{km}}}{\dfrac{p_\nu}{1\text{GeV}}} \tag{2.12}$$

This very useful formula gives the sensitivity of an experiment. An experiment usually aims to be sensitive to small δM^2. The ratio x/p_ν therefore has to be as big as possible. The best we can do is to use low energy neutrinos which travel a very long distance. For example, the IMB experiment has studied neutrinos created in the atmosphere by cosmic rays. It compares the flux of neutrinos born just above the detector with that of neutrinos which are born on the opposite side of the earth, and which must then traverse the whole earth ($x \sim 10^4$ km) to reach the detector (see Figure 3). This comparison gives a sensitivity to $\delta M^2 \sim 10^{-4}$ eV^2 for reasonably large mixing angles.[13]

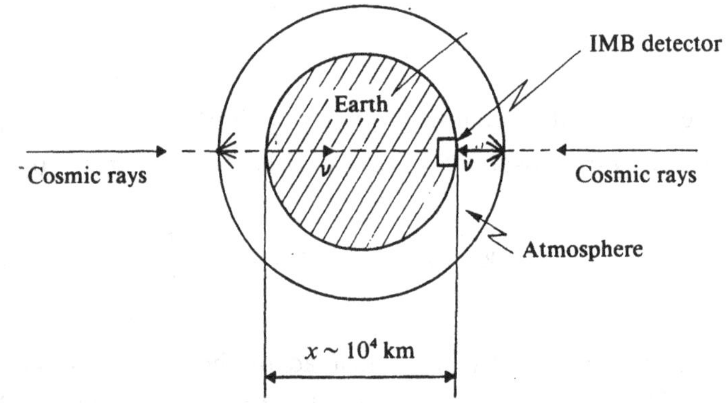

Fig. 3. An experiment to observe the oscillation of atmospheric neutrinos.

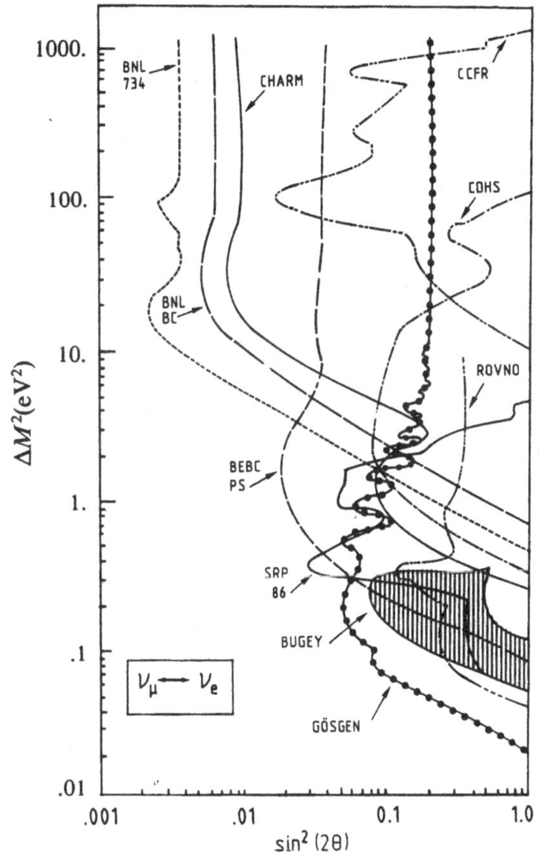

Fig. 4a. Limits in the plane (δM^2, $\sin^2 2\theta$) for the channel $\nu_\mu \leftrightarrow \nu_e$. (The shaded area corresponds to the region allowed by the BUGEY experiment.)

Almost all the experiments performed up to now have given negative results, and limits in the plane δM^2 versus $\sin^2 2\theta$ have been extracted. (Figure 4 a, b, c show those limits for three different channels, using data from both appearance and disappearance experiments.[14])

iv) Special oscillations

Apart from the oscillations we have already considered, there is also the possibility that a known neutrino, like ν_μ, could oscillate into a so-far

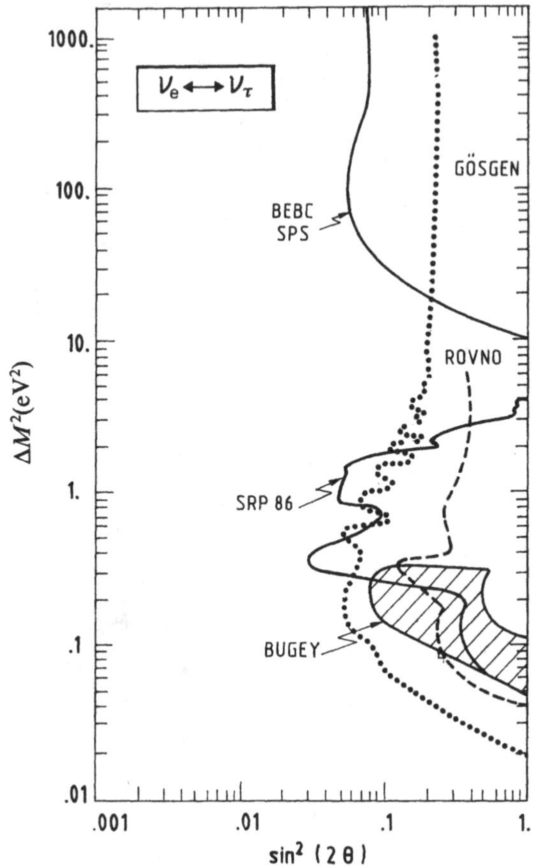

Fig. 4b. Limits in the plane $(\delta M^2, \sin^2 2\theta)$ for the channel $\nu_e \leftrightarrow \nu_\tau$. (The shaded area corresponds to the region allowed by the BUGEY experiment.)

unknown neutral lepton. For example, suppose there is a fourth lepton generation, containing a heavy charged lepton Φ, and an associated light neutrino ν_Φ. If a ν_μ whose energy is below the Φ production threshold oscillates into a ν_Φ, the ν_Φ will be unable to participate in charged-current interactions. It will be able to participate in neutral-current interactions, but if the detector observes only charged-current interactions, it will appear as if the ν_μ has turned into a sterile particle.

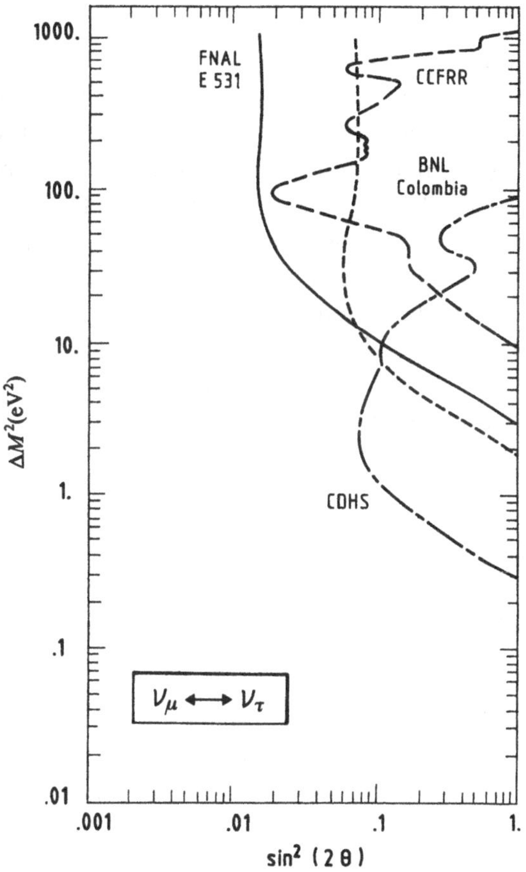

Fig. 4c. Limits in the plane (δM^2, $\sin^2 2\theta$) for the channel $\nu_\mu \leftrightarrow \nu_\tau$.

In addition, a known particle such as a ν_μ may oscillate into a truly sterile neutrino ν_s which does not participate in either the ordinary charged-current or neutral-current interactions at any energy. If this occurs, some of the flux in a ν_μ beam will simply disappear, without reappearing as ν_e or ν_τ flux.

As these two examples show, a disappearance oscillation experiment does not necessarily measure the same thing as an appearance experiment.

If neutrinos are Majorana particles, it is also possible for a ν_μ, interacting with matter, to produce a μ^+ or an e^+, rather than a μ^-. Although this effect varies with distance, it can actually occur as soon as the ν_μ is created. The point is that there is a small probability, of order $(M_m/E_m{}^*)^2$, for a given ν_m component of the ν_μ to be produced with positive helicity. Here $E_m{}^*$ is the energy of the ν_m in the rest frame of the particle (a pion, say) whose decay produces the neutrino. Now, in the Majorana case, a positive-helicity "neutrino" is what we normally call an antineutrino, and it will produce a positively charged lepton when interacting with matter. This leads to phenomena such as the one pictured below. The trouble is that $(M_m/E_m{}^*)^2$ is likely to be extremely tiny, so that the effect cannot be observed.

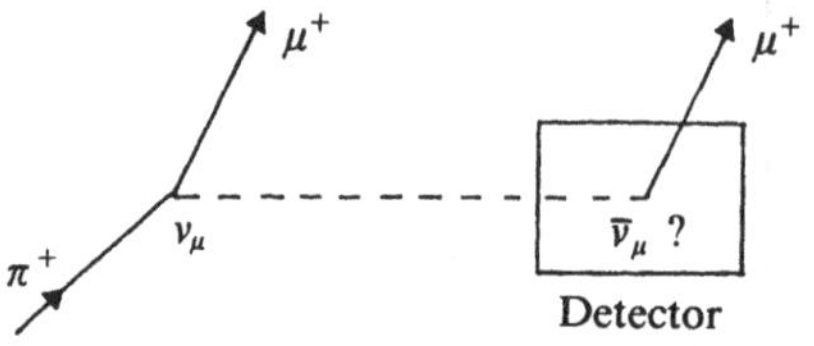

Up to now, we have discussed oscillation in vacuum. However, Mikheyev and Smirnov,[15] following earlier work of Wolfenstein,[16] have found recently a very exciting phenomenon which can occur when neutrinos travel through a dense material, and we will say a few things about it.

b) Neutrino flavor transitions in matter

Transitions from one neutrino flavor to another may have a very important connection to the famous solar neutrino mystery. Let us recall what this mystery is: the flux of electron neutrinos from the sun has been measured experimentally using the reaction:

$$\nu_e + {}^{37}\mathrm{Cl} \to {}^{37}\mathrm{Ar} + e^-$$

These neutrinos are coming from nuclear reactions occurring in the core of the sun, and the flux can be computed using the standard model of this star. The trouble is that the measured flux is a factor 4 too small compared to the calculation.

One idea to solve this discrepancy is to introduce neutrino masses and mixing between different neutrino species. Oscillation can then occur, and the ν_e flux on the earth is reduced. The distance between the source and the detector is very large, presumably a large multiple of the oscillation length, so that (see Eq. (2.9) and subsequent discussion) the reduction is given by:

$$P_{ee} = \sum_m |U_{em}|^4$$

If we accept the likely hypothesis that there is a mass eigenstate very "close" to ν_e, we find:

$$P_{ee} \sim 1$$

This means that the neutrino oscillation in vacuum is unlikely to be the explanation for the factor-of-four reduction of the solar ν_e flux. If we want to keep the phenomenon of neutrino flavor transition as an explanation, it seems that we have to assume that the mixing angles are very large. However, this is actually not necessary. If the neutrinos propagate through matter, the flavor transition pattern can be changed and one can get a very large reduction of the flux even with a small mixing angle. How does it work?

Let us take for example a two neutrino oscillation $\nu_e - \nu_\mu$. In vacuum we have:

$$\begin{aligned} \nu_e &= \quad \nu_1 \cos\theta_v + \nu_2 \sin\theta_v \\ \nu_\mu &= -\nu_1 \sin\theta_v + \nu_2 \cos\theta_v \end{aligned}$$

where ν_1 and ν_2 are the mass eigenstates, which propagate in vacuum without mixing, and θ_v is the mixing angle in vacuum.

If these neutrinos travel through the sun, they will interact with electrons and the propagation of ν_1 and ν_2 will be modified by the fact that ν_e and ν_μ have different interactions with electrons.

A ν_e interacts with an electron via both W and $Z°$ exchange,

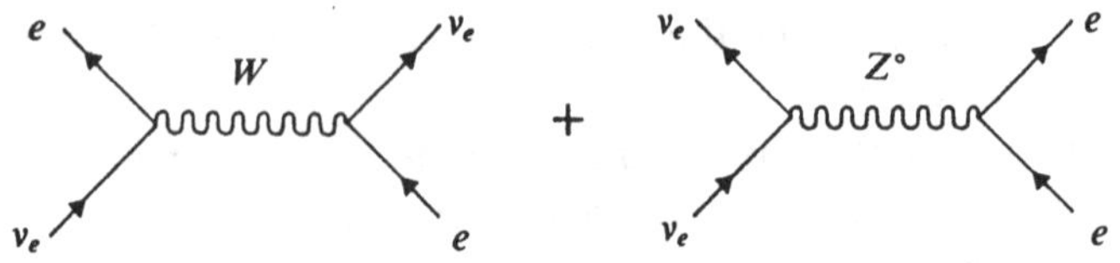

while a ν_μ (or a ν_τ) interacts only via $Z°$ exchange,

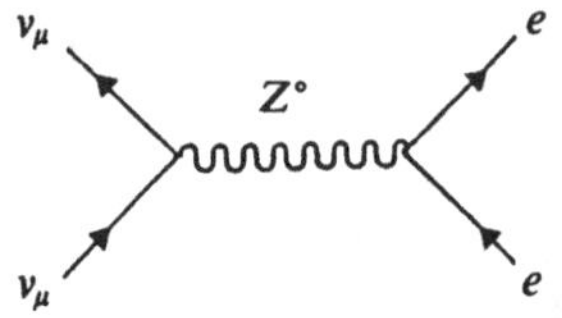

As a result of this difference, the eigenstates which propagate in solar matter without mixing (the "solar eigenstates") differ from those which propagate without mixing in vacuum (the mass eigenstates). The mixing angle θ_M in the solar matter (the rotation angle between $\nu_e - \nu_\mu$ and the solar eigenstates) is related to the mixing angle θ_v in vacuum (the rotation angle between $\nu_e - \nu_\mu$ and the mass eigenstates) by:

$$\sin^2 2\theta_M = \frac{\sin^2 2\theta_v}{\left(\cos 2\theta_v - \frac{L_v}{L_o}\right)^2 + \sin^2 2\theta_v} \tag{2.13}$$

Here L_v is the oscillation length in vacuum, and L_o, a length characteristic of the motion of the neutrinos through the solar matter, is defined by:

$$L_o = \frac{2\pi}{\sqrt{2}\, G_F N_e}$$

G_F is the Fermi constant

N_e is the number of electrons per unit volume in the sun.

Mikheyev and Smirnov have pointed out that Eq. (2.13) has a resonant character. Suppose that on their way from the solar core, where they are produced, to the outer edge of the sun, the solar neutrinos encounter a region where the electron density N_e is such that

$$\frac{L_v}{L_o} = \cos 2\theta_v$$

Then, according to Eq. (2.13), the mixing angle θ_M will be maximal,

$$\sin^2 2\theta_M = 1$$

even if the vacuum mixing angle θ_v is very small. Thus, very appreciable $\nu_e \to \nu_\mu$ conversion may occur in the sun, even for small θ_v. This Mikheyev-Smirnov-Wolfenstein conversion may be the explanation of the solar neutrino mystery. Interestingly, the values of δM^2 which can lead to the ν_e flux reduction observed in the ^{37}Cl solar neutrino experiment lie in the range 10^{-4} eV^2 to 10^{-8} eV^2.[17] Thus, future study of solar neutrinos may be a means to probe neutrino masses well below 1 eV.

EXERCISES

1 – We have derived the probability of neutrino oscillation, $P(\ell \to \ell', x)$, by assuming that a neutrino born as $\nu_\ell = \sum_m U_{\ell m}\nu_m$ has a well-defined momentum p_ν, so that its various mass eigenstate components ν_m have different energies $E_m = \sqrt{p_\nu^2 + M_m^2}$. Assume, instead, that this neutrino has a well-defined *energy* E_ν, so that its various mass eigenstate components have different *momenta* $p_m = \sqrt{E_\nu^2 - M_m^2}$. Show that this alternative assumption leads to the same wave function $\psi(x, x)$, Eq. (2.5), as before, except that p_ν is replaced by E_ν. Thus, this assumption leads to the same oscillation pattern as before.

2 – For the ν_e flux from the sun to be reduced appreciably by neutrino oscillation in vacuum, neutrino mixing must be large. Assuming that the earth is many oscillation lengths from the sun, and that CP is conserved,

vacuum oscillation reduces the ν_e flux by the factor:

$$P_{ee} = \sum_{m=1}^{N} U_{em}^4$$

Show that, if there are N different types of neutrinos, the minimum possible value of P_{ee} is $1/N$. Show further that if N is odd (three, for example), then the orthogonality of U makes it impossible for the flux of more than one neutrino flavor to be reduced to the minimum value $1/N$.

3 – Let $N_e(r)$ be the density of electrons in the sun as a function of the radius r, and let Δr be the thickness of a critical region, $r_{\text{crit}} - \frac{\Delta r}{2} < r < r_{\text{crit}} + \frac{\Delta r}{2}$, in which $N_e(r)$ is such that for neutrinos of a certain energy the solar mixing parameter $\sin^2 2\theta_M$ given by Eq. (2.13) exceeds, say, 1/2. If we are to have a very high degree of $\nu_e \to \nu_\mu$ conversion in this region by the Mikheyev-Smirnov-Wolfenstein mechanism, then Δr must be large enough to give the neutrinos ample time to convert before they reach a radius where $\sin^2 2\theta_M$ has become small again. Thus, we must have $\Delta r > L_M$, where L_M is the oscillation length of the neutrinos in the material of the critical region. Using the fact that $L_M = L_v/\sin 2\theta_v$,[16] show that the condition $\Delta r > L_M$ requires that:

$$\tan^2 2\theta_v \gtrsim \left[-\frac{1}{G_F N_e^2} \frac{dN_e}{dr} \right]_{r=r_{\text{crit}}}$$

Fortunately, this "adiabaticity constraint" on θ_v can be satisfied by quite small values.

CHAPTER

THREE

PHYSICS OF MASSIVE NEUTRAL LEPTONS

Having spent some time discussing the phenomenology, we would like now to discuss more extensively the physics of massive neutral leptons and in particular the physics of objects that are their own antiparticles, which is not very familiar. We shall study the properties of Majorana particles quite in detail. First we shall see that we can deduce a lot of interesting properties without using any formalism. Then, we shall discuss the formalism and write down the quantum field of a Majorana particle.

1 – PHYSICS WITHOUT THE FORMALISM OF FIELD THEORY

a) An approach to Majorana and Dirac particles

Here we shall make use only of basic quantum mechanics and discrete symmetries. We shall take our time to review *C*, *P* and *T* operators and then define precisely what we call a Majorana neutrino.

Suppose there exists a massive neutrino with negative helicity ν_-. However, our theory is CPT invariant: this means that there exists also the CPT mirror image of ν_-, an antineutrino with positive helicity: $\bar{\nu}_+$.

$$(\nu_- \quad , \quad \bar{\nu}_+)$$

$$\underbrace{\qquad\qquad\qquad}_{\text{CPT}}$$

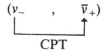

But our neutrino ν_- has a mass: its velocity is smaller than the speed of light so an observer can move faster. Then, in the frame of this observer, the neutrino is running the other way around but it is still spinning the same way. In other words, we have converted with a Lorentz boost our neutrino with negative helicity into a neutrino with positive helicity ν_+.

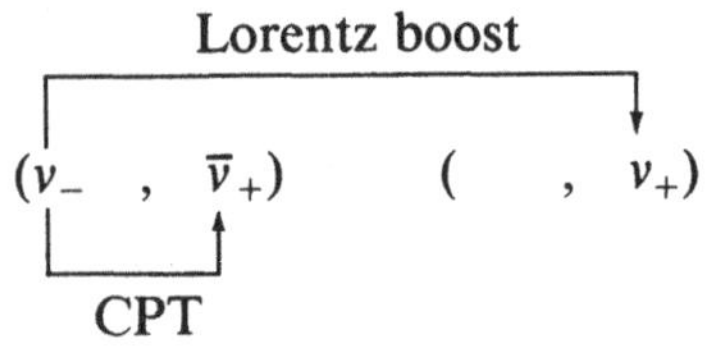

Now the question is: are these two states with positive helicity ν_+ and $\bar{\nu}_+$ the same? (That is, do they interact in the same way?).

• If we make the assumption that ν_+ *is not* the same particle as $\bar{\nu}_+$, then ν_+ has its own CPT mirror image $\bar{\nu}_-$. This new state can be connected to $\bar{\nu}_+$ by a Lorentz transformation.

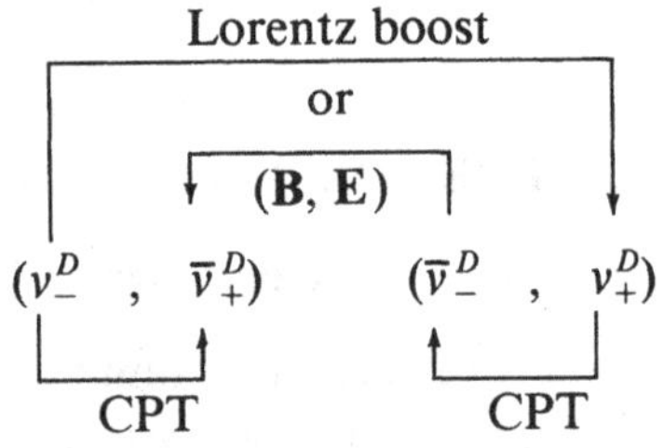

We have got four states with the same mass. *This set of states is called a Dirac neutrino* ν^D.

In general, a Dirac neutrino will have a magnetic dipole moment and an electric dipole moment. The two states $(\nu_-^D, \bar{\nu}_+^D)$ can then be converted into their partners $(\nu_+^D, \bar{\nu}_-^D)$ either by a Lorentz transformation or by the action of an external electromagnetic field ***B***, ***E***.

• If ν_+ is identical to $\bar{\nu}_+$, we are left with only two states with the same mass.

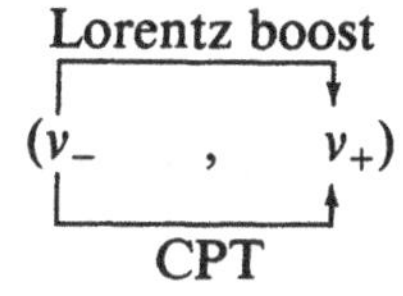

This set of two states is called a Majorana neutrino ν^M. In the rest frame, a CPT transformation applied to either of the two spin states simply reverses its spin. Then a 180° rotation brings the neutrino back to the original state.

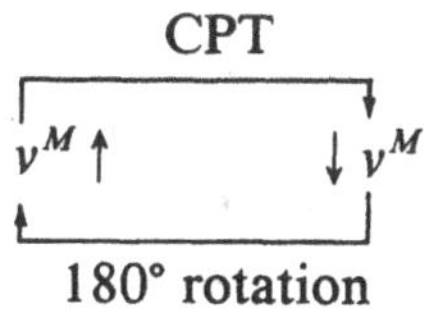

That is what we mean by saying that a Majorana neutrino is its own antiparticle: it goes back to itself under the action of CPT transformation followed by a 180° rotation. Before discussing in more detail the peculiar behaviour of Majorana neutrinos under C, P, and T, let us recall the main properties of these transformations.

b) Properties of *C*, *P*, *T*, transformations

i) Charge conjugation

The charge conjugation operator C is defined so that a one particle state $|f(\mathbf{p}, s)\rangle$ describing the particle f with momentum $\mathbf{p}$ and spin projection s is transformed into the one particle state $|\bar{f}(\mathbf{p}, s)\rangle$ describing the antiparticle $\bar{f}$ with the same spin projection and the same momentum.

$$C|f(\mathbf{p}, s)\rangle = \tilde{\eta}_C|\bar{f}(\mathbf{p}, s)\rangle \tag{3.1}$$

where $\tilde{\eta}_C$ is a phase factor.

ii) Parity

Under parity P, the space vector $\mathbf{r}$ is transformed into its opposite $-\mathbf{r}$.

The momentum $\mathbf{p}$ which is $m(d\mathbf{r}/dt)$ is then transformed into $-\mathbf{p}$ and the angular momentum $\mathbf{L} = \mathbf{r} \times \mathbf{p}$ is left invariant. For consistency, it is decided that every angular momentum $\mathbf{J}$ is left invariant. Thus we write:

$$P|f(\mathbf{p}, s)\rangle = \tilde{\eta}_P|f(-\mathbf{p}, s)\rangle \tag{3.2}$$

where $\tilde{\eta}_P$ is a phase factor which is often called the intrinsic parity of the particle f.

iii) Time reversal

The operation T consists of reversing the clock: t is then changed to $-t$ while $\mathbf{r}$ is kept invariant. The momentum $\mathbf{p} = m(d\mathbf{r}/dt)$ is thus changed to $-\mathbf{p}$ and the angular momentum $\mathbf{L} = \mathbf{r} \times \mathbf{p}$ is changed to $-\mathbf{L}$. Every angular momentum $\mathbf{J}$ is then changed to $-\mathbf{J}$. The action of time reversal on a one particle state $|f(\mathbf{p}, s)\rangle$ is described by an operator T such that:

$$T|f(\mathbf{p}, s)\rangle = \tilde{\eta}_T^s|f(-\mathbf{p}, -s)\rangle \tag{3.3}$$

where $\tilde{\eta}_T^s$ is a phase factor which depends on the initial spin s as we shall see later.

The operator T has a very special property: it is *antiunitary*. In order to understand what this means, let us come back to the old Schrödinger equation. In this framework, we expect from a useful symmetry to transform a solution of the Schrödinger equation into another solution of the Schrödinger equation.

Let us first try to define the time reversed wave function of $\psi(\mathbf{r}, t)$ by:

$$T\,\psi(\mathbf{r}, t) = \psi(\mathbf{r}, -t)$$

Unfortunately, this doesn't work: if $\psi(\mathbf{r}, t)$ is a solution of the Schrödinger equation, $\psi(\mathbf{r}, -t)$ is not. This can be easily shown using a plane wave function:

$$\psi(\mathbf{r}, t) = e^{i(\mathbf{p}\cdot\mathbf{r} - Et)}$$

This wave function is a solution of the Schrödinger equation

$$i\frac{\partial\psi}{\partial t} = -\frac{\nabla^2}{2m}\psi \qquad \text{if } E = \frac{\mathbf{p}^2}{2m}$$

Then, $T\,\psi(\mathbf{r}, t) = \exp\, i(\mathbf{p}{\cdot}\mathbf{r} + Et)$ is a solution of the Schrödinger equation if $E = -\mathbf{p}^2/2m$, but this cannot be the wave function of a physical object.

Wigner understood that the solution to this problem is to define:

$$T\,\psi(\mathbf{r}, t) = \psi^*(\mathbf{r}, -t) \tag{3.4}$$

The time reversed wave function of the plane wave is then:

$$T\,\psi(\mathbf{r}, t) = e^{i(-\mathbf{p}\cdot\mathbf{r} - Et)}$$

which is now a solution of the Schrödinger equation with $E = \mathbf{p}^2/2m$ describing the same particle going in the opposite direction as expected.

Some unusual properties follow from the definition (3.4) above:

(i)
$$T\,[\alpha\,|f(\mathbf{p}, s)\rangle] = \alpha^* T\,|f(\mathbf{p}, s)\rangle \tag{3.5}$$

T is said to be *antilinear.*

(ii) Let us consider the scalar product $\langle T\varphi\,|\,T\psi\rangle$ where φ and ψ are two time independent states:

$$\begin{aligned}\langle T\varphi\,|\,T\psi\rangle &= \int d^3r\,(T\varphi)^*\,(T\psi)\\ &= \int d^3r\,\varphi\psi^*\\ &= \langle\psi\,|\,\varphi\rangle\end{aligned}$$

We know that if one has an operator U such that:

$$\langle U\varphi\,|\,U\psi\rangle = \langle\varphi\,|\,\psi\rangle$$

then U is a unitary operator. One can also write:

$$\langle U\varphi | U\psi \rangle = \langle \varphi | U^{\dagger} U\psi \rangle = \langle \varphi | \psi \rangle$$

where $U^{\dagger}$ is the adjoint of U.

In our case, something different happens:

$$\langle T\varphi | T\psi \rangle = \langle \psi | \varphi \rangle \tag{3.6}$$

T is said to be *antiunitary*.

As a consequence of this, the operator CPT is also antiunitary. Before coming back to Majorana neutrinos, let us add that when using antiunitary operators, one has to be very careful; most of the usual rules do not hold.

c) C, P, T properties of Majorana neutrinos

i) CP properties and phases

As a first statement, we would like to stress that a physical Majorana neutrino is not an eigenstate of charge conjugation.

$$C | \nu^M(\mathbf{p}, s) \rangle \neq \tilde{\eta}_C | \nu^M(\mathbf{p}, s) \rangle \tag{3.7}$$

In free field theory, a Majorana neutrino will be defined to be an eigenstate of C. This property is no longer maintained when you switch on the weak interaction. Let us indeed consider the two diagrams of Figure 5.

These two amplitudes are the C mirror images of each other and they contribute to the physical propagator. However, the weak interaction is left-handed. Therefore, the second amplitude is very much suppressed compared to the first one, since the current which creates the intermediate ℓ^+ couples mostly to *positive* helicity neutrinos. A Majorana neutrino is not an eigenstate of C because it has maximally C-violating interactions.

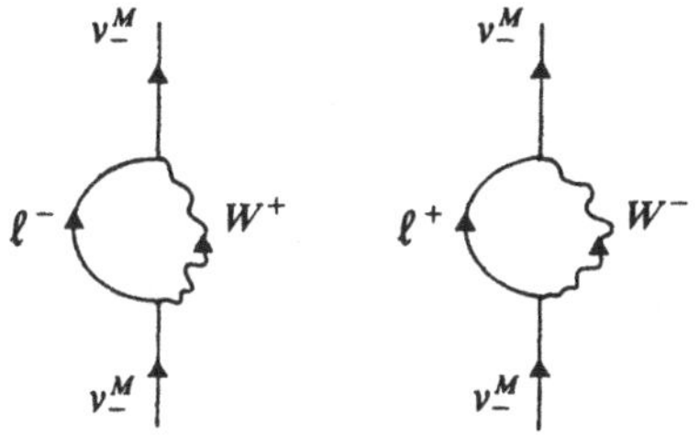

Figure 5. A Majorana neutrino is not an eigenstate of C.

The photino, which is a Majorana object also, is an eigenstate of C: its interactions don't violate C to a good approximation.

The weak interaction violates CP also but if we neglect this presumably tiny effect, a Majorana neutrino is an eigenstate of CP:

$$\mathrm{CP}\,|\,\nu^M(\mathbf{p}, s)\rangle = \tilde{\eta}_{\mathrm{CP}}\,|\,\nu^M(-\mathbf{p}, s)\rangle$$

This has immediate consequences:

Let us for example suppose that there exists a heavy neutral lepton N which is a Majorana particle and which decays into electron and hadrons through CP conserving interactions. Then the decay rates $N \to e^+ +$ hadrons and $N \to e^- +$ hadrons are very simply related.

N is a Majorana particle, therefore:

$$\mathrm{CP}\,|\,N(\mathbf{p}, s)\rangle = \text{phase}\,|\,N(-\mathbf{p}, s)\rangle$$

Let $\Gamma(N \to e^+\text{had})$ be the decay rate of N into $e^+ +$ hadrons and let us choose the rest frame of N.

$$\Gamma(N \to e^+\text{had}) = \int |\langle e^+(\mathbf{p}_e), \text{had}(\mathbf{p}_x)\,|\,H\,|\,N(s)\rangle|^2$$

where H is the CP conserving Hamiltonian,

$$(\mathrm{CP})\,H\,(\mathrm{CP})^{-1} = H$$

and the integration is over all outgoing momenta and spins and includes phase space factors.

So:

$$\Gamma(N \to e^+\text{had}) = \int |\langle e^+(\mathbf{p}_e), \text{had}(\mathbf{p}_x) | (\text{CP})^{-1}(\text{CP})H(\text{CP})^{-1}(\text{CP}) | N(s) \rangle|^2$$

$$= \int |\langle e^-(-\mathbf{p}_e), \overline{\text{had}}(-\mathbf{p}_x) | H | N(s) \rangle|^2$$

(all phases disappear)

$\overline{\text{had}}$ is the CP mirror image of the hadronic system had. This gives after integration over momentum and over the final spins which we have suppressed

$$\Gamma(N \to e^+\text{had}) = \Gamma(N \to e^-\overline{\text{had}})$$

That is, the Majorana particle N not only decays to either an e^- or an e^+, but does so with equal rates![18] Therefore, if such a particle is produced, it will be very easy to tell whether it is a Majorana or Dirac particle. The Majorana character of N could be detected in deep inelastic e^-p scattering or in the decay of W bosons in $p\bar{p}$ collisions where the e^+ in the decay is detected, for example (Figure 6).

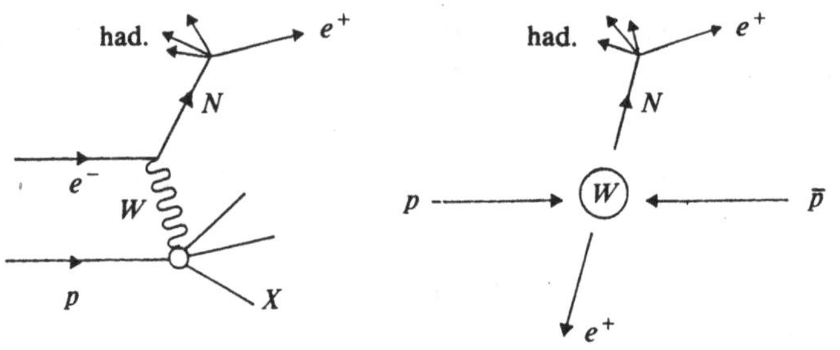

Figure 6. Signatures of a heavy Majorana lepton.

Let us now look in more detail at the phase factor $\tilde{\eta}_{CP}$, the intrinsic CP of ν^M:

We know that the intrinsic CP parity of the photon is $+1$; it is also $+1$ for the $Z°$ and it is -1 for the $\pi°$. What is the intrinsic CP parity of a Majorana neutrino ν^M?

In order to find this out, let us imagine the decay $Z° \to \nu^M \nu^M$ assuming that CP is conserved in this decay (which is true in the standard model) and that the final neutrinos are nonrelativistic. In the initial state, the $Z°$ has $J = 1$ so the possible final states are:

L	S	
1	0	$^1P_1$
0	1	$^3S_1$
1	1	$^3P_1$
2	1	$^3D_1$

But we have two identical fermions in the final state so the wave function must be antisymmetric: only $^3P_1$ satisfies this condition. The action of CP on the final state is then:

$$\mathrm{CP}\,|\,\nu^M\nu^M;\,{}^3P_1\rangle = \tilde{\eta}_{CP}{}^2(-1)^L\,|\,\nu^M\nu^M;\,{}^3P_1\rangle$$

The factor $(-1)^L$ is the analogue for a two-particle state of definite L of the momentum reversal in the one-particle state. Since $L = 1$ and the initial CP parity is $+1$, we derive:

$$+1 = -\tilde{\eta}_{CP}{}^2$$

Thus:

$$\boxed{\tilde{\eta}_{CP}(\nu^M) = \pm i} \qquad (3.8)$$

The intrinsic CP parity of a Majorana neutrino is imaginary. Note that two different Majorana neutrinos can have opposite CP parities; this is very important for the double β decay as we shall see later.

This imaginary CP parity has, at least in principle, practical consequences:

Consider $\pi^\circ \rightarrow \nu^M + \nu^M$

The initial state of π° is 0^{-+} so the possible final states are:

$$J = 0 \qquad S = 0 \qquad L = 0 \qquad {}^1S_0$$

$$J = 0 \qquad S = 1 \qquad L = 1 \qquad {}^3P_0$$

$$\mathrm{CP}\,|\,\nu^M\nu^M\rangle = (\pm\, i)^2\,(-1)^L\,|\,\nu^M\nu^M\rangle$$

If CP is conserved in this decay, since $\mathrm{CP}\,|\,\pi^\circ\rangle = -\,|\,\pi^\circ\rangle$, then we get:

$$-1 = (\pm i)^2\,(-1)^L$$

$$\therefore\; L = 0$$

Only the final state ${}^1S_0$ is allowed: the two neutrinos are emitted with their spins antiparallel. (Had $\tilde{\eta}_{\mathrm{CP}}\,(\nu^M)$ been real, the CP-allowed final state would have been the ${}^3P_0$, so that parallel neutrino spins would have been possible.)

ii) CPT properties and phases

From the discussion in Chapter III-1 (a and b), the effect of the CPT operator on a ν^M state is:

$$\mathrm{CPT}\,|\,\nu^M(\mathbf{p}, s)\rangle = \tilde{\eta}_{\mathrm{CPT}}{}^s\,|\,\nu^M(\mathbf{p}, -s)\rangle \tag{3.9}$$

Here the phase factor $\tilde{\eta}_{\mathrm{CPT}}{}^s$ depends on the initial spin state, as we now show.

From here on we use the notation: $\zeta = \mathrm{CPT}$.

In the rest frame of the neutrino we have:

$$\zeta\,|\,\nu^M(s)\rangle = \tilde{\eta}^s_\zeta\,|\,\nu^M(-s)\rangle$$

$$s = \pm\frac{1}{2}$$

This relation implies that, so long as we act only on the states $|\nu^M(s)\rangle$, ζ anticommutes with $\mathbf{J}$:

$$\zeta\mathbf{J} = -\mathbf{J}\zeta$$

Let $J_\pm$ be the raising and lowering operators:

$$J_\pm = J_x \pm iJ_y$$

Then

$$\begin{aligned} \zeta J_+ &= \zeta(J_x + iJ_y) \\ &= \zeta J_x - i\zeta J_y \qquad\qquad (\zeta \text{ is antilinear}) \\ &= -J_x\zeta + iJ_y\zeta \\ &= -J_-\zeta \end{aligned}$$

Let us apply this relation to the down-state $\left|-\tfrac{1}{2}\right\rangle$. We have

$$\begin{aligned} \zeta J_+\left|-\tfrac{1}{2}\right\rangle &= \zeta\left|+\tfrac{1}{2}\right\rangle = \tilde{\eta}_\zeta^{\;+1/2}\left|-\tfrac{1}{2}\right\rangle = \\ &= -J_-\zeta\left|-\tfrac{1}{2}\right\rangle = -J_-\,\tilde{\eta}_\zeta^{\;-1/2}\left|+\tfrac{1}{2}\right\rangle = -\tilde{\eta}_\zeta^{\;-1/2}\left|-\tfrac{1}{2}\right\rangle \end{aligned}$$

From this we see that:[19]

$$\boxed{\tilde{\eta}_\zeta^{\;+1/2} = -\tilde{\eta}_\zeta^{\;-1/2}} \tag{3.10}$$

Let us now derive some consequences of the CPT properties of Majorana particles.

d) Magnetic and electric dipole moments of Majorana neutrinos

Suppose we have a Majorana neutrino ν^M in a uniform, static electromagnetic field $\mathbf{B}$, $\mathbf{E}$. If this neutrino ν^M has a magnetic dipole moment μ_{mag} and an electric dipole moment μ_{el}, its interaction energy with the external electromagnetic field is:

$$E_{int} = -\mu_{mag}\langle \mathbf{s}\cdot\mathbf{B}\rangle - \mu_{el}\langle \mathbf{s}\cdot\mathbf{E}\rangle$$

What happens if we perform a CPT transformation? First of all, **s** is changed to $-\mathbf{s}$. Now what about **B** and **E**? The CPT properties of uniform, static electromagnetic fields are given in Table 1. **B** and **E** are odd under C (the intrinsic C parity of the photon is -1: if you reverse the sign of the electric charges, the signs of the fields are reversed) and are invariant under CPT.

As a result, under CPT, E_{int} goes into $-E_{int}$. This implies that, if the world is CPT invariant,

$$\mu_{mag} = 0 \qquad \text{and} \qquad \mu_{el} = 0$$

Majorana neutrinos have no magnetic dipole moment and no electric dipole moment. Dirac neutrinos may have dipole moments, but in the absence of right-handed currents they are proportional to the mass of the neutrino [20] and much too small to give rise to experimentally observable phenomena.[21, 22]

Table 1: C, P, T properties of the electromagnetic field.

	$\mathbf{A}$	φ	$\mathbf{B} = \nabla\times\mathbf{A}$	$\mathbf{E} = -\nabla\varphi - \partial\mathbf{A}/\partial t$
C	$-$	$-$	$-$	$-$
P	$-$	$+$	$+$	$-$
T	$-$	$+$	$-$	$+$
CPT	$-$	$-$	$+$	$+$

e) Coupling to the photon of a Majorana neutrino

Since a Majorana neutrino is not only electrically neutral but also has no dipole moments, we may well ask whether it can couple to a photon at all. The answer is yes. We can find the most general form for the matrix element $\langle \nu^M | J_\mu^{EM} | \nu^M \rangle$ of the electromagnetic current of a Majorana neutrino using the CPT property (3.9) and the CPT phase relation (3.10).[23,24,25] These imply (see exercises) that

$$\begin{aligned}\langle \nu^M(\mathbf{p}_f, s_f) | J_\mu^{EM} | \nu^M(\mathbf{p}_i, s_i) \rangle &= -\tilde{\eta}_\zeta^{s_i *} \tilde{\eta}_\zeta^{s_f} \langle \nu^M(\mathbf{p}_i, -s_i) | J_\mu^{EM} | \nu^M(\mathbf{p}_f, -s_f) \rangle \\ &= -(-1)^{s_i - s_f} \langle \nu^M(\mathbf{p}_i, -s_i) | J_\mu^{EM} | \nu^M(\mathbf{p}_f, -s_f) \rangle\end{aligned}$$

Here we have used the fact that the four-vector potential A_μ is CPT-odd (see Table 1), so that J_μ^{EM} must also be odd if the electromagnetic interaction $H^{EM} = J_\mu^{EM} A_\mu$ is to be CPT invariant. Now, for *any* spin-1/2 fermion, it follows from Lorentz invariance and current conservation that

$$\begin{aligned}\langle \mathbf{p}_f, s_f | J_\mu^{EM} | \mathbf{p}_i, s_i \rangle = i\, \bar{u}(\mathbf{p}_f, s_f) \,[F\gamma_\mu &+ G(q^2\gamma_\mu - \not{q} q_\mu)\gamma_5 \\ &+ M\sigma_{\mu\nu} q_\nu + E i \sigma_{\mu\nu} q_\nu \gamma_5] u(\mathbf{p}_i, s_i)\end{aligned}$$

Here u and $\bar{u}$ are Dirac spinors, $q = p_i - p_f$, and F, G, M, and E are form factors which depend on q^2. Using this relation on both sides of the CPT constraint we just wrote down for the case of a Majorana neutrino, it is not hard to show (see exercises) that for such a particle, F, M, and E must vanish. That is, a Majorana neutrino can have just the one form factor G:

$$\langle \nu^M(\mathbf{p}_f, s_f) | J_\mu^{EM} | \nu^M(\mathbf{p}_i, s_i) \rangle = i\bar{u}_f G(q^2)\,(q^2\gamma_\mu - \not{q} q_\mu)\gamma_5 u_i$$

To what physical electromagnetic structure inside the neutrino does this matrix element correspond? It is not obvious, but the answer is the following:[26] Imagine taking a solenoid and bending it so that the ends come together and it forms a doughnut. Join the windings at the ends to form one continuous winding. There is now a **B** field circulating around the inside of the doughnut.

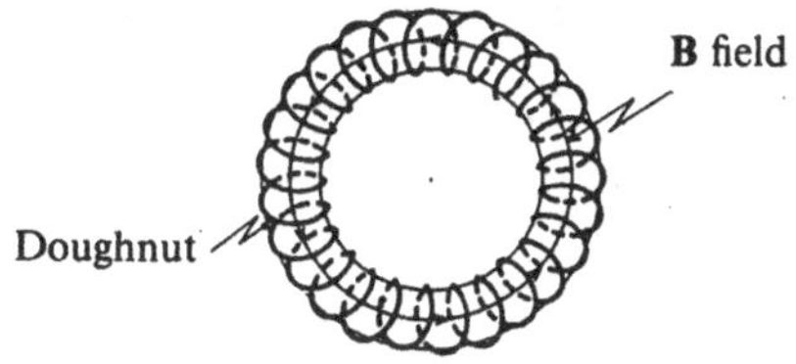

Such a **B** field is the possible electromagnetic structure of a Majorana neutrino. There is no **B** field at infinity (nor any **E** field, for that matter), consistent with the fact that a ν^M has no dipole moments. One can check (try it) that the structure sketched above behaves under CPT as it should if it is to be part of a Majorana neutrino.

2 - THE QUANTUM MAJORANA FIELD

We would like now to give an idea about the properties of the interactions of a Majorana particle and introduce very important notations and concepts used in the physics of massive neutrinos. We shall first review basic features of the Dirac equation, but the distinction between Dirac and Majorana particles will occur only at the quantum field level.

a) The Dirac equation

i) Notations, conventions

We will use the notations, the metric, and the γ matrices described in the book of Sakurai.[27]

The time component of the four vectors is $x_4 = ict$. The Dirac equation of a free fermion is:

$$\left[\gamma_\mu \frac{\partial}{\partial x_\mu} + M\right] \psi(x) = 0$$

For a given momentum **p**, this equation has four linearly independent solutions:

$$u(\mathbf{p}, s) = \sqrt{\frac{E+M}{2M}} \begin{bmatrix} 1 \\ \dfrac{\boldsymbol{\sigma}\cdot\mathbf{p}}{E+M} \end{bmatrix} \chi_s, \quad v(\mathbf{p}, s) = \sqrt{\frac{E+M}{2M}} \begin{bmatrix} \dfrac{\boldsymbol{\sigma}\cdot\mathbf{p}}{E+M} \\ 1 \end{bmatrix} \chi_s^c$$

$$s = \pm \frac{1}{2} \qquad (3.11)$$

with:

$$\chi_{s=1/2} = \begin{bmatrix} 1 \\ 0 \end{bmatrix} \qquad \chi^c_{s=1/2} = -\begin{bmatrix} 0 \\ 1 \end{bmatrix}$$

$$\chi_{s=-1/2} = \begin{bmatrix} 0 \\ 1 \end{bmatrix} \qquad \chi^c_{s=-1/2} = \begin{bmatrix} 1 \\ 0 \end{bmatrix} \tag{3.12}$$

These solutions satisfy: $(\ i\not{p} + M)u(\mathbf{p}, s) = 0$
$(-i\not{p} + M)v(\mathbf{p}, s) = 0$

where $\not{p} = \gamma_\mu p_\mu$ with $p_\mu = (\mathbf{p}, iE)$, $E > 0$.
Usually, $u(\mathbf{p}, s)$ describes particles and $v(\mathbf{p}, s)$ describes antiparticles.
The γ matrices are written in the Pauli-Dirac representation:

$$\gamma_i = \begin{bmatrix} 0 & -i\sigma_i \\ i\sigma_i & 0 \end{bmatrix} \qquad \gamma_4 = \begin{bmatrix} I & 0 \\ 0 & -I \end{bmatrix} \tag{3.13}$$

where σ_i are the usual Pauli matrices. The γ matrices have the following properties:

$$\gamma_\mu\gamma_\nu + \gamma_\nu\gamma_\mu = 2\delta_{\mu\nu}$$
$$\gamma_\mu^\dagger = \gamma_\mu$$

which means:

$$\gamma_\mu^t = -\gamma_\mu \qquad \mu = 1, 3$$
$$\gamma_\mu^t = \gamma_\mu \qquad \mu = 2, 4 \tag{3.14}$$

We define:

$$\gamma_5 = \gamma_1\gamma_2\gamma_3\gamma_4 = \begin{bmatrix} 0 & -I \\ -I & 0 \end{bmatrix} \tag{3.15}$$

γ_5 has the following properties:

$$\gamma_5\gamma_\mu + \gamma_\mu\gamma_5 = 0$$
$$\gamma_5^\dagger = \gamma_5, \qquad \gamma_5^2 = I \tag{3.16}$$

Some useful relations are collected in Table 2.

Table 2: Useful properties of spinors.

$$\gamma_2 u^*(\mathbf{p}, s) = v(\mathbf{p}, s)$$
$$\gamma_2 v^*(\mathbf{p}, s) = u(\mathbf{p}, s)$$
$$\gamma_5 v(\mathbf{p}, s) = (-1)^{s-1/2} u(\mathbf{p}, -s)$$
$$\gamma_5 u(\mathbf{p}, s) = -(-1)^{s-1/2} v(\mathbf{p}, -s)$$
$$\not{a}^2 = +a^2$$

Table 3: Properties of the charge conjugation matrix Ω.

$$\Omega^{-1}\gamma_\mu\Omega = -\gamma_\mu^t$$
$$\Omega^{-1} = \Omega^\dagger = \Omega^t = -\Omega = -\Omega^*$$

In particular: $\Omega\gamma_4 + \gamma_4\Omega = 0$

$$\Omega\gamma_5 - \gamma_5\Omega = 0$$
$$\psi^c = \Omega\bar{\psi}^t$$
$$\overline{\psi^c} = \psi^t\Omega = -\psi^t\Omega^{-1}$$
$$(\psi_{L,R})^c = (\psi^c)_{R,L}$$

ii) Charge conjugation in the Dirac equation

The Dirac equation of a charged fermion interacting with an external electromagnetic field is:

$$\left[\gamma_\mu\left(\frac{\partial}{\partial x_\mu} - ieA_\mu\right) + M\right]\psi(x) = 0$$

We would like to find the spinor ψ^c describing a fermion of charge $-e$ interacting with the field A_μ. The spinor ψ^c satisfies:

$$\left[\gamma_\mu\left(\frac{\partial}{\partial x_\mu} + ieA_\mu\right) + M\right]\psi^c = 0 \qquad (3.17)$$

We try:[28]

$$\psi^c = S_c\,\psi^*$$

where S_c is a 4×4 matrix, and ψ^* is the column vector with $(\psi^*)_\alpha = (\psi_\alpha)^*$. Then we have:

$$\left[\gamma_\mu\left(\frac{\partial}{\partial x_\mu} + ieA_\mu\right) + M\right] S_c\psi^* = 0$$

Complex conjugating this equation and multiplying by $(S_c^*)^{-1}$, we have

$$\left[(S_c^*)^{-1}\gamma_k{}^*S_c^*\left(\frac{\partial}{\partial x_k} - ie\,A_k\right) - (S_c^*)^{-1}\gamma_4{}^*S_c^*\left(\frac{\partial}{\partial x_4} - ie\,A_4\right) + M\right]\psi = 0$$

Comparing with the original Dirac equation, we see that ψ^c will satisfy Eq. (3.17) if

$$(S_c^*)^{-1}\gamma_k{}^*S_c^* = \gamma_k \quad \text{and} \quad (S_c^*)^{-1}\gamma_4{}^*S_c^* = -\gamma_4\,.$$

It is easy to see that in our representation of the gamma matrices,

$$S_c = \gamma_2$$

fulfills these requirements. Thus,

$$\psi^c = \gamma_2\,\psi^* = \Omega\,\bar{\psi}^t \tag{3.18}$$

where

$$\Omega = \gamma_2\,\gamma_4 \tag{3.19}$$

The spinor ψ^c is said to be the *charge conjugate* of the wave function ψ, and Ω is called the *charge conjugation matrix*. We note that

$$\Omega^{-1} = \Omega^\dagger = \Omega^t = -\Omega = -\Omega^* \tag{3.20}$$

and that Ω satisfies

$$\Omega^{-1}\gamma_\mu \Omega = \gamma'_\mu \tag{3.21}$$

Other properties of the charge conjugation matrix are collected in Table 3.

b) Charge conjugation for quantum fermion fields

The free Dirac field is given by the following plane-wave expansion:

$$\Psi^D(x) = \sum_{\mathbf{p},s} \sqrt{\frac{M}{E(\mathbf{p})V}}\,[f(\mathbf{p},s)u(\mathbf{p},s)e^{ipx} + \bar{f}^\dagger(\mathbf{p},s)v(\mathbf{p},s)e^{-ipx}] \tag{3.22}$$

where:
$f(\mathbf{p}, s)$ is the annihilation operator for the particle f
$\bar{f}^\dagger(\mathbf{p}, s)$ is the creation operator for the antiparticle $\bar{f}$
V is some volume associated to the normalisation of the states.

We would like to find the properties of this quantum field under charge conjugation.

The charge conjugate of a fermionic field $\Psi(x)$ is:

$$C\Psi(x)C^{-1} \tag{3.23}$$

where C is a unitary charge conjugation operator. For a Ψ which is either a Dirac or a Majorana field, $C\Psi\, C^{-1}$ is defined by:

$$C\Psi(x)C^{-1} = \eta_C^*\,\gamma_2\Psi^*(x) \tag{3.24}$$

where $\Psi^*(x)$ is the column vector:

$$\begin{bmatrix} \Psi_1^\dagger(x) \\ \Psi_2^\dagger(x) \\ \Psi_3^\dagger(x) \\ \Psi_4^\dagger(x) \end{bmatrix}$$

if $\Psi_i(x)$ are the componants of $\Psi(x)$. η_C is a phase factor.

QED is invariant under this operation when it is combined with a related transformation of the photon field.

Let us show that this definition of the charge conjugate field has all the expected properties when Ψ is a Dirac field Ψ^D. First, we note that:

$$[\Psi^D(x)]^* = \sum_{\mathbf{p},s} \sqrt{\frac{M}{E(\mathbf{p})V}} \, [f^\dagger(\mathbf{p}, s) u^*(\mathbf{p}, s) e^{-ipx} + \bar{f}(\mathbf{p}, s) v^*(\mathbf{p}, s) e^{ipx}]$$

so that:

$$\eta_C^* \gamma_2 [\Psi^D(x)]^* = \eta_C^* \sum_{\mathbf{p},s} \sqrt{\frac{M}{E(\mathbf{p})V}} \, [f^\dagger(\mathbf{p}, s) \gamma_2 u^*(\mathbf{p}, s) e^{-ipx} + \bar{f}(\mathbf{p}, s) \gamma_2 v^*(\mathbf{p}, s) e^{ipx}]$$

Now:

$$\gamma_2 u^*(\mathbf{p}, s) = v(\mathbf{p}, s)$$
$$\gamma_2 v^*(\mathbf{p}, s) = u(\mathbf{p}, s)$$

Hence:

$$\eta_C^* \gamma_2 [\Psi^D(x)]^* = \eta_C^* \sum_{\mathbf{p},s} \sqrt{\frac{M}{E(\mathbf{p})V}} \, [\bar{f}(\mathbf{p}, s) u(\mathbf{p}, s) e^{ipx} + f^\dagger(\mathbf{p}, s) v(\mathbf{p}, s) e^{-ipx}]$$

One can see that, apart from the phase η_C^*, $C \Psi^D C^{-1} = \eta_C^* \gamma_2 [\Psi^D]^*$ is nothing but the field Ψ^D with the operators $f(\mathbf{p}, s)$ and $\bar{f}(\mathbf{p}, s)$ interchanged. In the literature, one sometimes encounters the quantity

$$\Psi^c(x) \equiv \gamma_2 \Psi^*(x)$$

where Ψ may be either a Dirac or Majorana field. Ψ^c is, of course, just

$C\Psi C^{-1}$ without the phase η_C^*, and we have just shown that for a Dirac field:

$$[\Psi^D]^c = \Psi^D\,(f(\mathbf{p}, s) \leftrightarrow \bar{f}(\mathbf{p}, s))$$

This interchange of particle and antiparticle is exactly what one expects to happen under charge conjugation.

Now, since C is a linear operator, applying it to the plane-wave expansion of Ψ^D gives

$$\begin{aligned} C\Psi^D C^{-1} &= C\sum_{\mathbf{p},s}\sqrt{\frac{M}{E(\mathbf{p})V}}\,[f(\mathbf{p}, s)u(\mathbf{p}, s)e^{ipx} + \bar{f}^\dagger(\mathbf{p}, s)v(\mathbf{p}, s)e^{-ipx}]C^{-1} \\ &= \sum_{\mathbf{p},s}\sqrt{\frac{M}{E(\mathbf{p})V}}\,[Cf(\mathbf{p}, s)C^{-1}u(\mathbf{p}, s)e^{ipx} \\ &\quad + C\bar{f}^\dagger(\mathbf{p}, s)C^{-1}v(\mathbf{p}, s)e^{-ipx}] \end{aligned}$$

Comparing this with the plane-wave expansion of $\eta_C^*\,\gamma_2\,[\Psi^D]^*$, we find that

$$Cf(\mathbf{p}, s)C^{-1} = \eta_C^*\,\bar{f}(\mathbf{p}, s)$$

Taking the hermitian conjugate gives (remember that $C^{-1} = C^\dagger$):

$$Cf^\dagger(\mathbf{p}, s)\,C^{-1} = \eta_C\bar{f}^\dagger(\mathbf{p}, s)$$

Let us apply this relation to the vacuum $|0\rangle$. We define C so that:

$$C|0\rangle = |0\rangle$$

Then we have:

$$Cf^\dagger(\mathbf{p}, s)|0\rangle = \eta_C\bar{f}^\dagger(\mathbf{p}, s)|0\rangle$$

which gives for the one-particle states:

$$C|f(\mathbf{p}, s)\rangle = \eta_C|\bar{f}(\mathbf{p}, s)\rangle$$

Thus, our definition of C in terms of its effect on fields, Eq. (3.24), implies that the states transform under C as one would expect, and as we have previously stated in Eq. (3.1). Further, we see that the phase $\tilde{\eta}_C$ defined by Eq. (3.1) is just η_C.

c) The quantum Majorana field

In the case of the Majorana fermion, there is no difference between particle and antiparticle. One might then guess that the plane wave expansion of Ψ^M is simply:

$$\Psi^M(x) = \sum_{\mathbf{p}, s} \sqrt{\frac{M}{E(\mathbf{p})V}} \left[f(\mathbf{p}, s)u(\mathbf{p}, s)e^{ipx} + f^\dagger(\mathbf{p}, s)v(\mathbf{p}, s)^{-ipx}\right]$$

This guess is actually correct apart from a little detail: $\bar{f}$ and f are in fact not exactly the same, but can differ by a phase, and we have:

$$\Psi^M(x) = \sum_{\mathbf{p}, s} \sqrt{\frac{M}{E(\mathbf{p})V}} \left[f(\mathbf{p}, s)u(\mathbf{p}, s)e^{ipx} + \lambda f^\dagger(\mathbf{p}, s)v(\mathbf{p}, s)^{-ipx}\right] \tag{3.25}$$

where λ is a phase factor which is called the *creation phase factor.* [19]

i) Origin of the creation phase factor

We would like to explain now why we need to introduce λ. Let us see what happens if we redefine the phases of the one-particle states. This we may always do if we wish. The new one-particle state is:

$$|f'(\mathbf{p}, s)\rangle = e^{i\varphi}|f(\mathbf{p}, s)\rangle$$

$$\text{but:} \quad |f'(\mathbf{p}, s)\rangle = f'^\dagger(\mathbf{p}, s)|0\rangle$$

$$|f(\mathbf{p}, s)\rangle = f^\dagger(\mathbf{p}, s)|0\rangle$$

$$\text{so:} \qquad f'^{\dagger}(\mathbf{p}, s) = e^{i\varphi} f^{\dagger}(\mathbf{p}, s)$$
$$f'(\mathbf{p}, s) = e^{-i\varphi} f(\mathbf{p}, s)$$

The conventional field Ψ^M annihilates the one-particle state with a specific matrix element:

$$\langle 0 | \Psi^M(0) | f(\mathbf{p}, s) \rangle = \sqrt{\frac{M}{E(\mathbf{p})V}}\, u(\mathbf{p}, s)$$

The new field Ψ'^M that annihilates the new state with the same matrix element is:

$$\Psi'^M = e^{-i\varphi} \Psi^M$$

Using the expansion of Ψ^M:

$$\Psi'^M(x) = e^{-i\varphi} \sum_{\mathbf{p}, s} \sqrt{\frac{M}{E(\mathbf{p})V}} \left[f(\mathbf{p}, s) u(\mathbf{p}, s) e^{ipx} + \lambda f^{\dagger}(\mathbf{p}, s) v(\mathbf{p}, s) e^{-ipx} \right]$$
$$= \sum_{\mathbf{p}, s} \sqrt{\frac{M}{E(\mathbf{p})V}} \left[f'(\mathbf{p}, s) u(\mathbf{p}, s) e^{ipx} + \lambda e^{-2i\varphi} f'^{\dagger}(\mathbf{p}, s) v(\mathbf{p}, s) e^{-ipx} \right]$$

which can be written:

$$\Psi'^M(x) = \sum_{\mathbf{p}, s} \sqrt{\frac{M}{E(\mathbf{p})V}} \left[f'(\mathbf{p}, s) u(\mathbf{p}, s) e^{ipx} + \lambda' f'^{\dagger}(\mathbf{p}, s) v(\mathbf{p}, s) e^{-ipx} \right]$$

with:

$$\lambda' = \lambda\, e^{-2i\varphi}$$

Therefore λ is an arbitrary phase factor which depends on the set of one-particle states you choose. The most convenient choice, however, is often not $\lambda = 1$ because this does not lead to a real mixing matrix U when CP is conserved. We shall come back to that later.

ii) CPT properties of the quantum Majorana field

Let us check that the field Ψ^M actually describes Majorana objects as we have defined them before.

For any fermion field Ψ, the action of the CPT operator is given by:

$$\zeta\Psi(x)\zeta^{-1} = -\eta_\zeta^*\gamma_5\Psi^*(-x) \tag{3.26}$$

In the case of the Majorana field we get:

$$\begin{aligned}\zeta\Psi^M(x)\zeta^{-1} &= \zeta\sum_{\mathbf{p},s}\sqrt{\frac{M}{E(\mathbf{p})V}}\,[f(\mathbf{p},s)u(\mathbf{p},s)e^{ipx} \\ &\quad + \lambda f^\dagger(\mathbf{p},s)v(\mathbf{p},s)e^{-ipx}]\zeta^{-1} \\ &= \sum_{\mathbf{p},s}\sqrt{\frac{M}{E(\mathbf{p})V}}\,[\zeta f(\mathbf{p},s)\zeta^{-1}u^*(\mathbf{p},s)e^{-ipx} \\ &\quad + \lambda^*\zeta f^\dagger(\mathbf{p},s)\zeta^{-1}v^*(\mathbf{p},s)e^{ipx}]\end{aligned}$$

because ζ is antilinear.
On the other hand we have:

$$\begin{aligned}-\eta_\zeta^*\gamma_5\Psi^*(-x) &= -\eta_\zeta^*\gamma_5\sum_{\mathbf{p},s}\sqrt{\frac{M}{E(\mathbf{p})V}} \\ &\quad [f^\dagger(\mathbf{p},s)u^*(\mathbf{p},s)e^{ipx} + \lambda^* f(\mathbf{p},s)v^*(\mathbf{p},s)e^{-ipx}]\end{aligned}$$

Now: $\gamma_5\, v(\mathbf{p},s) = (-1)^{s-1/2}\,u(\mathbf{p},-s)$

(see Table 2)

$\gamma_5\, u(\mathbf{p},s) = -(-1)^{s-1/2}\,v(\mathbf{p},-s)$

and $\gamma_5 = \gamma_5^*$, so the quantity above is:

$$= \eta_\zeta^*\sum_{\mathbf{p},s}\sqrt{\frac{M}{E(\mathbf{p})V}}\,(-1)^{s-1/2}$$

$$[\lambda^* f(\mathbf{p}, -s) u^*(\mathbf{p}, s) e^{-ipx} - f^\dagger(\mathbf{p}, -s) v^*(\mathbf{p}, s) e^{ipx}]$$

Identifying $\zeta \Psi^M \zeta^{-1}$ with this last expression we get:

$$\zeta f(\mathbf{p}, s) \zeta^{-1} = \eta_\zeta^* \lambda^* (-1)^{s-1/2} f(\mathbf{p}, -s)$$
$$\zeta f^\dagger(\mathbf{p}, s) \zeta^{-1} = -\eta_\zeta^* \lambda (-1)^{s-1/2} f^\dagger(\mathbf{p}, -s)$$

To compare these two equations, we have to use a trick because the adjoint $\zeta^\dagger$ of the antiunitary operator ζ is not ζ^{-1}. Therefore, we proceed by noting that for any operator Q:

$$\langle b | Q | a \rangle = \langle \zeta a | \zeta Q^\dagger \zeta^{-1} | \zeta b \rangle$$

Thus, if we define $\zeta | 0 \rangle = | 0 \rangle$, we have:

$$\begin{aligned} 1 &= \langle 0 | f(\mathbf{p}, s) f^\dagger(\mathbf{p}, s) | 0 \rangle \\ &= \langle 0 | \zeta f(\mathbf{p}, s) \zeta^{-1} \zeta f^\dagger(\mathbf{p}, s) \zeta^{-1} | 0 \rangle \end{aligned}$$

which gives, if we use the relations above:

$$1 = -\eta_\zeta^{*2} \tag{3.27}$$

The phase factor η_ζ is imaginary.

If we now apply the equation for $\zeta f^\dagger(\mathbf{p}, s) \zeta^{-1}$ to the vacuum, we obtain:

$$\zeta | f(\mathbf{p}, s) \rangle = \eta_\zeta \lambda (-1)^{s-1/2} | f(\mathbf{p}, -s) \rangle$$

Thus, the quanta of the field Ψ^M of Eq. (3.25) really are Majorana particles as we have defined them before. That is, under CPT they go into themselves, except for a reversal of the spin, and the picking up of a phase factor (cf. Eq. (3.9)). We see that this phase factor $\tilde{\eta}_\zeta^s$, defined by Eq. (3.9), is given in terms of η_ζ and λ by:

$$\tilde{\eta}_\zeta^s = \eta_\zeta \lambda (-1)^{s-1/2} \tag{3.28}$$

This result conforms to the rule $\tilde{\eta}_\zeta^{+1/2} = -\tilde{\eta}_\zeta^{-1/2}$ which we proved before without using field theory.

iii) C properties of the quantum Majorana field

Now, apply the charge conjugation to Ψ^M:

$$C\Psi^M C^{-1} = \eta_C^* \gamma_2 (\Psi^M)^* \qquad \text{by definition.}$$

$$\begin{aligned}
\eta_C^* \gamma_2 (\Psi^M)^* &= \eta_C^* \sum_{\mathbf{p},s} \sqrt{\frac{M}{E(\mathbf{p})V}} [f^\dagger(\mathbf{p}, s)\gamma_2 u^*(\mathbf{p}, s) e^{-ipx} \\
&\quad + \lambda^* f(\mathbf{p}, s)\gamma_2 v^*(\mathbf{p}, s) e^{ipx}] \\
&= \eta_C^* \sum_{\mathbf{p},s} \sqrt{\frac{M}{E(\mathbf{p})V}} [f^\dagger(\mathbf{p}, s) v(\mathbf{p}, s) e^{-ipx} + \lambda^* f(\mathbf{p}, s) u(\mathbf{p}, s) e^{ipx}] \\
&= \eta_C^* \lambda^* \Psi^M
\end{aligned}$$

so that:

$$\boxed{C\Psi^M C^{-1} = \lambda^* \eta_C^* \Psi^M} \tag{3.29}$$

Apart from phase, a Majorana field is its own charge conjugate. Inserting the plane-wave expansion of Ψ^M into $C\Psi^M C^{-1}$, one derives easily:

$$C f(\mathbf{p}, s) C^{-1} = \eta_C^* \lambda^* f(\mathbf{p}, s)$$

$$C f^\dagger(\mathbf{p}, s) C^{-1} = \eta_C^* \lambda^* f^\dagger(\mathbf{p}, s)$$

The consistency of these two relations requires that $\eta_C \lambda$ be real.

Applying the second relation to the vacuum as before then gives:

$$C|f(\mathbf{p}, s)\rangle = \tilde{\eta}_C|f(\mathbf{p}, s)\rangle$$

with

$$\boxed{\tilde{\eta}_C = \lambda \eta_C} \tag{3.30}$$

In free field theory, the Majorana particle is an eigenstate of C. Furthermore, we have demonstrated that: if Ψ^M is a Majorana field, the conjugate field $C\,\Psi^M C^{-1}$ is the original field Ψ^M times a phase factor. Now consider the field Ψ^D, Eq. (3.22), and note that for suitable $\bar{f}$, this field is actually completely general, and describes either the Dirac or the Majorana situation. Suppose that the charge conjugate $C\,\Psi\,C^{-1}$ of some fermion field Ψ of the form given in Eq. (3.22) is equal to Ψ, apart from a phase factor. Remembering that $C\,\Psi\,C^{-1}/\eta_C^*$ is just Ψ with $f \leftrightarrow \bar{f}$, we see that this would mean that $\bar{f}$ = (phase) f. Thus, Ψ must be a Majorana field and describe a Majorana particle. We have thus proved that:

Ψ is a Majorana field if and only if $C\,\Psi\,C^{-1}$ = (phase)$\cdot\Psi$

We will give in the next chapter some examples to show how the Majorana field is handled in practice.

d) Helicity rules for Majorana neutrinos

In developing a qualitative picture of processes involving Majorana neutrinos, it is useful to know how the helicities of these neutrinos behave when they are emitted or absorbed by weak currents of definite handedness. In terms of the left- and right-handed chiral projection operators

$$P_L = \frac{1 + \gamma_5}{2}$$

and

$$P_R = \frac{1 - \gamma_5}{2} \tag{3.31}$$

the charged weak currents of handedness $H(= L$ or $R)$ are of the form $\overline{\ell}\,\gamma_\mu P_H \nu$ or its hermitean conjugate $\overline{\nu}\gamma_\mu P_H \ell$. In these expressions, ℓ is some charged lepton, and the neutrino field ν may be either of Dirac or Majorana character.

As we have seen, the Dirac neutrino field has the structure:

$$\nu^D \sim \Sigma\,[fu + \bar{f}^\dagger \upsilon] \qquad \text{and} \qquad \overline{\nu^D} \sim \Sigma\,[f^\dagger \bar{u} + \bar{f}\,\bar{\upsilon}]$$

Correspondingly, the Majorana neutrino field has the structure:

$$\nu^M \sim \Sigma[fu + \lambda f^\dagger \upsilon] \qquad \text{and} \qquad \overline{\nu^M} \sim \Sigma[f^\dagger\,\bar{u} + \lambda^* f \bar{\upsilon}]$$

From these expressions, it follows that when a weak current of handedness H acts at some vertex, the amplitude for the vertex will contain the factor given in Table 4 below. It is important to notice that, as a consequence of the structure of the neutrino fields, *this factor does not depend on whether neutrinos are Dirac or Majorana particles.* Thus, the helicity properties of neutrinos emitted or absorbed at weak vertices do not depend on this distinction either.

Table 4: Factor in the amplitude for a weak vertex at which a current of handedness H acts.

Charged lepton emitted or absorbed	Current that acts	Neutrino (or antineutrino) emitted or absorbed	Factor in amplitude
ℓ^- emitted or ℓ^+ absorbed	$\overline{\ell}\gamma_\mu P_H \nu$	Absorbed	$P_H u$
		Emitted	$P_H \upsilon$
ℓ^- absorbed or ℓ^+ emitted	$\overline{\nu}\gamma_\mu P_H \ell$	Emitted	$\bar{u}\gamma_\mu P_H = \overline{P_H u}\gamma_\mu$
		Absorbed	$\bar{\upsilon}\gamma_\mu P_H = \overline{P_H \upsilon}\gamma_\mu$

To see what helicity properties the amplitude factors in Table 4 imply, let us recall that for given momentum $\mathbf{p}$ and helicity Λ $(\equiv \pm 1)$, the

Dirac spinors satisfy:

$$\begin{aligned} Q_\Lambda\, u(\mathbf{p}, \Lambda) &= u(\mathbf{p}, \Lambda) \quad , \quad Q_\Lambda\, u(\mathbf{p}, -\Lambda) = 0 \\ Q_\Lambda\, \upsilon(\mathbf{p}, \Lambda) &= \upsilon(\mathbf{p}, \Lambda) \quad , \quad Q_\Lambda\, \upsilon(\mathbf{p}, -\Lambda) = 0 \end{aligned} \tag{3.32}$$

Here Q_Λ is the helicity projection operator:

$$Q_\Lambda = \frac{1 + i\gamma_5 \Lambda \not{w}}{2}$$

where
$$w_\mu = \frac{1}{M}\left(E\frac{\mathbf{p}}{|\mathbf{p}|}, i|\mathbf{p}|\right) \tag{3.33}$$

When the neutrino is highly relativistic, so that $E \simeq |\mathbf{p}|$,

$$w_\mu \cong \frac{1}{M}(\mathbf{p}, iE) = \frac{1}{M}p_\mu$$

Replacing $\not{w}$ by $\not{p}/M$ in Q_Λ, Eq. (3.33), and referring to the Dirac equations satisfied by u and υ, we see that for highly relativistic particles:

$$Q_\Lambda \cong \begin{cases} \dfrac{1 - \Lambda\gamma_5}{2} & \text{when acting on } u \\[2ex] \dfrac{1 + \Lambda\gamma_5}{2} & \text{when acting on } \upsilon \end{cases} \tag{3.34}$$

That is, the projection operator for right-handed (left-handed) *helicity* reduces to the right-handed (left-handed) *chiral* projection operator when acting on u, and to the left-handed (right-handed) *chiral* projection operator when acting on υ. In view of this reduction, the neutrinos associated with a factor $P_H u$ or $\overline{P_H u}$ in Table 4 will have helicity of handedness H, while those associated with a factor $P_H \upsilon$ or $\overline{P_H \upsilon}$ will have helicity of handedness opposite to H.

For the Dirac case, we see from the structure of ν^D and $\overline{\nu^D}$ that the neutrinos are always described by a u (or $\bar{u}$) and the antineutrinos by a υ

(or $\bar{\upsilon}$). Thus, in the relativistic limit, a Dirac neutrino can couple to a left-handed weak current only if it has left-handed helicity, and a Dirac antineutrino only if it has right-handed helicity. A similar situation holds for the coupling to a right-handed weak current.

For the Majorana case, the neutrino helicity may be read off from Table 4. However, it is easier simply to note, as we already have, that the helicity will be *as if* the neutrino were a Dirac particle. Table 4 shows that the type of spinor, u or υ, that is associated with a given neutrino depends on the charge of the ℓ to which the neutrino is coupled, on whether the ℓ is emitted or absorbed, and on whether the neutrino is emitted or absorbed. Furthermore, the dependence is such that in the Dirac case, a neutrino is always associated with a u, and an antineutrino with a υ. Thus, to determine the helicity of a *Majorana* neutrino coupled to some weak vertex, one can pretend that it is a Dirac neutrino, and ask whether one would call this "Dirac neutrino" a ν^D or a $\overline{\nu^D}$, given the details of the vertex. If one would call it a ν^D, its helicity will have the same handedness as the current at the vertex. If one would call it a $\overline{\nu^D}$, its helicity will have opposite handedness from that of the current.

To the extent that they have mass, neutrinos are not fully relativistic, and the helicity rules we have discussed can be violated. The amplitude for a neutrino of mass M and energy E to be emitted or absorbed by a weak current in spite of having a helicity opposite to that dictated by the rules above is of order M/E. To illustrate this, let us consider a $\overline{\nu^D}$ or ν^M emitted by a left-handed current in the decay $W^- \to \ell^- + \overset{(-)}{\nu}$. This neutrino will be described by a spinor υ. According to the rules, it must be right-handed. Indeed, from Eqs. (3.11) and (3.12):

$$\sqrt{\frac{M}{E}}\,\upsilon(\mathbf{p},\Lambda) \xrightarrow[M\to 0]{} \frac{1}{\sqrt{2}}\begin{bmatrix} -\Lambda \\ 1 \end{bmatrix} \chi^c_\Lambda$$

and, since:

$$P_L = \frac{1+\gamma_5}{2} = \frac{1}{2}\begin{bmatrix} 1 & -1 \\ -1 & 1 \end{bmatrix}$$

we have:

$$P_L\left[\frac{1}{\sqrt{2}}\begin{bmatrix}-\Lambda\\1\end{bmatrix}\chi_\Lambda^c\right]=\begin{cases}\left[\frac{1}{\sqrt{2}}\begin{bmatrix}-\Lambda\\1\end{bmatrix}\chi_\Lambda^c\right], & \Lambda=+1\\ 0, & \Lambda=-1\end{cases}$$

However, it is easy to show that for small, non-vanishing M/E,

$$P_L\left[\sqrt{\frac{M}{E}}\,v(\mathbf{p},\Lambda=-1)\right]\cong\frac{1}{2}\frac{M}{E}\left[\frac{1}{\sqrt{2}}\begin{bmatrix}-1\\1\end{bmatrix}\chi_\Lambda^c\right]$$

Thus, relative to the normal weak amplitude, the amplitude for the neutrino to be emitted with the "wrong" (i.e., left-handed) helicity is of order M/E, as claimed.

EXERCISES

1 – Suppose $\tilde{\gamma}$ is the spin-1/2 superpartner of the γ in supersymmetric theories. In such theories, $\tilde{\gamma}$ is a Majorana particle. Assuming that the process $\gamma+\gamma\to\tilde{\gamma}+\tilde{\gamma}$ conserves C, show that $\tilde{\eta}_C(\tilde{\gamma})$, the intrinsic C parity of the $\tilde{\gamma}$, is ± 1. This result implies that if $\tilde{\eta}_{CP}(\tilde{\gamma})=\pm i$ as for a Majorana neutrino, (as will be shown to be true in exercise III.5), then $\tilde{\eta}_P(\tilde{\gamma})=\pm i$ as well. Here $\tilde{\eta}_P(\tilde{\gamma})$ is the intrinsic parity of the $\tilde{\gamma}$.

2 – Suppose that the masses of the toponium states and of the $\tilde{\gamma}$ are such that the $^3P_2$ state of toponium can decay into a pair of nonrelativistic $\tilde{\gamma}$'s. Show, using the imaginary character of $\tilde{\eta}_P(\tilde{\gamma})$, that if P (parity) is conserved in this decay, the $\tilde{\gamma}$'s must be either in a $^3P_2$ state or in a $^3F_2$ state. What additional state is allowed if P is violated? Note that because P conservation forbids this additional state, P conservation (and, in particular, the fact that $\tilde{\eta}_P(\tilde{\gamma})$ is imaginary) influences the $\tilde{\gamma}\tilde{\gamma}$ angular distribution in the decay.

3 – Using the CPT property (3.9), the CPT phase relation (3.10), and the

fact that the electromagnetic current J_μ^{EM} is CPT-odd, show that:

$$\begin{aligned}\langle \nu^M(\mathbf{p}_f, s_f) | J_\mu^{EM} | \nu^M(\mathbf{p}_i, s_i)\rangle &= -\tilde{\eta}_\zeta^{s_i *} \tilde{\eta}_\zeta^{s_f} \langle \nu^M(\mathbf{p}_i, -s_i) | J_\mu^{EM} | \nu^M(\mathbf{p}_f, -s_f)\rangle \\ &= -(-1)^{s_i - s_f} \langle \nu^M(\mathbf{p}_i, -s_i) | J_\mu^{EM} | \nu^M(\mathbf{p}_f, -s_f)\rangle\end{aligned}$$

4 – For an arbitrary spin-1/2 fermion f:

$$\begin{aligned}\langle f(\mathbf{p}_f, s_f) | J_\mu^{EM} | f(\mathbf{p}_i, s_i)\rangle = i\, \bar{u}(\mathbf{p}_f, s_f)\, [F\gamma_\mu + G(q^2\gamma_\mu - \not{q} q_\mu)\gamma_5 \\ + M\sigma_{\mu\nu} q_\nu + Ei\sigma_{\mu\nu} q_\nu \gamma_5]\, u(\mathbf{p}_i, s_i)\end{aligned}$$

Here $q = p_i - p_f$, and F, G, M, and E are form factors which depend on q^2. Using this relation on both sides of the CPT constraint proved in problem III.3, show that for a Majorana neutrino F, M, and E vanish. In doing this it is useful to know that:

$$u(\mathbf{p}, -s) = (-1)^{s-1/2}\, \gamma_5\, \Omega\, \bar{u}^t(p, s)$$

as follows from Table 2. It is also useful to know that:

$$\Omega^{-1}\, \Gamma\, \Omega = +\Gamma^t \text{ for } \Gamma = 1, \gamma_5, \gamma_\mu \gamma_5$$

and

$$\Omega^{-1}\, \Gamma\, \Omega = -\Gamma^t \text{ for } \Gamma = \gamma_\mu, \sigma_{\mu\nu}, \sigma_{\mu\nu}\gamma_5$$

5 – For any fermion field Ψ, the effect of a CP transformation is defined by:

$$\mathrm{CP}\, \Psi(\mathbf{x}, t)\, (\mathrm{CP})^{-1} = \eta_{\mathrm{CP}}^{*}\, \gamma_4\, \gamma_2\, \Psi^{*}(-\mathbf{x}, t)$$

where η_{CP} is a phase factor. This definition implies a corresponding CP transformation law for the state $|f\rangle$ which Ψ annihilates. Show that, depending on whether Ψ and $|f\rangle$ are a Dirac field and particle or a

Majorana field and particle, the implied transformation law for $|f\rangle$ is:

Dirac case: $\quad \mathrm{CP}\,|f(\mathbf{p}, s)\rangle = \eta_{\mathrm{CP}}\,|\bar{f}(-\mathbf{p}, s)\rangle$

Majorana case: $\quad \mathrm{CP}\,|f(\mathbf{p}, s)\rangle = \tilde{\eta}_{\mathrm{CP}}\,|f(-\mathbf{p}, s)\rangle$

with $\tilde{\eta}_{\mathrm{CP}} = \eta_{\mathrm{CP}}\,\lambda = -\tilde{\eta}^{*}_{\mathrm{CP}}$.

In the Majorana case, λ is the creation phase factor in Ψ. Note that the constraint $\tilde{\eta}_{\mathrm{CP}} = -\tilde{\eta}^{*}_{\mathrm{CP}}$ states that the intrinsic CP parity of any Majorana particle must be imaginary.

In carrying out this exercise, it is important to know that:[27] $\gamma_4\, u(\mathbf{p}, s) = u(-\mathbf{p}, s)$, $\gamma_4\, v(\mathbf{p}, s) = -v(-\mathbf{p}, s)$.

CHAPTER

FOUR

PROCESSES WITH MAJORANA FIELDS

1 - NEUTRAL CURRENT NEUTRINO INTERACTIONS

Here we shall explain why, despite the enormous amount of neutral current neutrino scattering data, we have not been able to tell if the neutrino is a Majorana or a Dirac particle.

Consider a neutral current neutrino scattering:

$$\nu + A \rightarrow \nu + B$$

In the standard model, this process is described by a Hamiltonian density:

$$H \sim (\nu NC)_\mu M_\mu(A, B)$$

where $(\nu NC)_\mu = \bar{\nu} \gamma_\mu (1 + \gamma_5) \nu$ is a purely left-handed current coupled to $M_\mu(A, B)$, which is some operator pertaining to A and B.

The vector part of the neutrino current is: $\bar{\nu}\gamma_\mu \nu$. What is $\overline{\nu^c}\gamma_\mu \nu^c$?

Remember that for any fermion field Ψ, Ψ^c is defined as $\gamma_2 \Psi^*$, the exact field-theory analogue of the charge-conjugate wave function Ψ^c in Dirac theory. Thus, from Table 3 we have $\Psi^c = \gamma_2\Psi^* = \Omega\bar{\Psi}^t$ and $\overline{\psi^c} = -\Psi^t\Omega^{-1}$.

Then: $$\overline{\nu^c}\gamma_\mu \nu^c = -\nu^t \Omega^{-1}\gamma_\mu \Omega \bar{\nu}^t$$

But we have the property (see Table 3):

$$\Omega^{-1}\gamma_\mu \Omega = -\gamma_\mu^t$$

Thus, our quantity of interest is

$$\begin{aligned}\overline{\nu^c}\gamma_\mu \nu^c &= \nu^t \gamma_\mu^t \bar{\nu}^t \\ &= \nu_\alpha (\gamma_\mu)_{\beta\alpha}\, \bar{\nu}_\beta \\ &= -\bar{\nu}_\beta (\gamma_\mu)_{\beta\alpha}\, \nu_\alpha\end{aligned}$$

because fermion fields anticommute.
Hence:

$$\overline{\nu^c}\gamma_\mu \nu^c = -\bar{\nu}\gamma_\mu \nu \tag{4.1}$$

This is true for any fermion, Dirac or Majorana (the vector current of the positron is opposite to the vector current of the electron).

For a Majorana field, we have the additional condition:

$$(\nu^M)^c = \text{phase}\cdot\nu^M$$

so: $$\overline{(\nu^M)^c}\gamma_\mu (\nu^M)^c = \overline{\nu^M}\gamma_\mu \nu^M \tag{4.2}$$

In view of Eq. (4.1), this means $\overline{\nu^M}\gamma_\mu \nu^M = 0$.

A Majorana particle has no vector current.

This is not really a surprise. For example, the fermion number operator is:

$$\int \bar{\nu}\gamma_4 \nu d^3x = N_\nu - N_{\bar{\nu}} \qquad \text{for any fermion}$$

This is, of course, zero for a Majorana neutrino.

The only contribution of a Majorana neutrino to the neutral current scattering is then through the axial vector current.

Let us compute the matrix element:

$$\langle \nu_f^M | \overline{\nu^M} \gamma_\mu \gamma_5 \nu^M | \nu_i^M \rangle$$

We have:

$$\Psi^M(x) = \sum_{\mathbf{p},s} \sqrt{\frac{M}{E(\mathbf{p})V}} \left[f(\mathbf{p},s) u(\mathbf{p},s) e^{ipx} + \lambda f^\dagger(\mathbf{p},s) v(\mathbf{p},s) e^{-ipx} \right]$$

$$\bar{\Psi}^M(x) = \sum_{\mathbf{p},s} \sqrt{\frac{M}{E(\mathbf{p})V}} \left[\lambda^* f(\mathbf{p},s) \bar{v}(\mathbf{p},s) e^{ipx} + f^\dagger(\mathbf{p},s) \bar{u}(\mathbf{p},s) e^{-ipx} \right]$$

Inserting these expressions in the current gives (omitting irrelevant normalization factors):

$$\begin{aligned} \langle \nu_f^M | \overline{\nu^M} \gamma_\mu \gamma_5 \nu^M | \nu_i^M \rangle &= \sum_{\mathbf{p},s} \sum_{\mathbf{p}',s'} \langle \nu_f^M | [\lambda^* f(\mathbf{p},s) \bar{v}(\mathbf{p},s) e^{ipx} \\ &\quad + f^\dagger(\mathbf{p},s) \bar{u}(\mathbf{p},s) e^{-ipx}] \gamma_\mu \gamma_5 [f(\mathbf{p}',s') u(\mathbf{p}',s') e^{ip'x} \\ &\quad + \lambda f^\dagger(\mathbf{p}',s') v(\mathbf{p}',s') e^{-ip'x}] | \nu_i^M \rangle \end{aligned}$$

But we have:

$$\langle \nu_f^M | f(\mathbf{p},s) f(\mathbf{p}',s') | \nu_i^M \rangle = 0$$

$$\langle \nu_f^M | f^\dagger(\mathbf{p},s) f^\dagger(\mathbf{p}',s') | \nu_i^M \rangle = 0$$

$$\langle \nu_f^M | f^\dagger(\mathbf{p},s) f(\mathbf{p}',s') | \nu_i^M \rangle = \delta_{s',s_i} \delta_{s,s_f} \delta(\mathbf{p}_i - \mathbf{p}') \delta(\mathbf{p} - \mathbf{p}_f)$$

$$\langle \nu_f^M | f(\mathbf{p},s) f^\dagger(\mathbf{p}',s') | \nu_i^M \rangle = -\delta_{s',s_f} \delta_{s,s_i} \delta(\mathbf{p}_i - \mathbf{p}) \delta(\mathbf{p}' - \mathbf{p}_f)$$

using the commutation relations between fermion fields:

$$[f(\mathbf{p},s), f^\dagger(\mathbf{p}',s')] \sim \delta_{ss'} \delta(\mathbf{p} - \mathbf{p}')$$

Thus, we obtain:

$$\langle \nu_f^M | \left[\overline{\nu^M}\, \gamma_\mu \gamma_5\, \nu^M \right]_{x=0} | \nu_i^M \rangle = \bar{u}_f \gamma_\mu \gamma_5\, u_i - \bar{v}_i\, \gamma_\mu \gamma_5\, v_f$$

This expression can be simplified. We have already used (see Table 2):

$$\gamma_2 u^*(\mathbf{p}, s) = v(\mathbf{p}, s)$$

One can rewrite this as:

$$v(\mathbf{p}, s) = u^c(\mathbf{p}, s) = \Omega\, \bar{u}^t(\mathbf{p}, s)$$

which implies that:

$$\bar{v}(\mathbf{p}, s) = -u^t(\mathbf{p}, s)\, \Omega^{-1}$$

Then:

$$\bar{v}_i \gamma_\mu \gamma_5 v_f = -u_i^t\, \Omega^{-1}\, \gamma_\mu \gamma_5 \Omega\, \bar{u}_f^t$$

But: γ_5 commutes with Ω

$$\Omega^{-1} \gamma_\mu \Omega = -\gamma_\mu^t$$

and $\gamma_5^t = \gamma_5$

so: $$\Omega^{-1}\, \gamma_\mu \gamma_5\, \Omega = -\gamma_\mu^t \gamma_5^t = -(\gamma_5\, \gamma_\mu)^t = +\, (\gamma_\mu\, \gamma_5)^t$$

(an axial vector current is even under C).

Then: $$\bar{v}_i\, \gamma_\mu\, \gamma_5\, v_f = -u_i^t\, (\gamma_\mu\, \gamma_5)^t \bar{u}_f^t$$

therefore: $$\bar{v}_i \gamma_\mu \gamma_5 v_f = -\bar{u}_f \gamma_\mu \gamma_5 u_i$$

and: $$\langle \nu_f^M | \overline{\nu^M}\, \gamma_\mu \gamma_5 \nu^M | \nu_i^M \rangle = 2\, \bar{u}_f \gamma_\mu \gamma_5 u_i \tag{4.3}$$

For a Dirac neutrino, we would get:

$$\langle \nu_f^D | \overline{\nu^D} \gamma_\mu (1 + \gamma_5) \nu^D | \nu_i^D \rangle = \bar{u}_f \gamma_\mu (1 + \gamma_5) u_i \qquad (4.4)$$

Why is there no difference experimentally between these two cases?

It comes from the fact that our neutrino beam is left-handed and relativistic. The spinor u_i for a neutrino in such a beam obeys

$$\gamma_5 u_i = u_i + O\left(\frac{M}{E}\right)$$

Hence:

$$2\bar{u}_f \gamma_\mu \gamma_5 u_i \xrightarrow[\frac{M}{E} \to 0]{} \bar{u}_f \gamma_\mu (1 + \gamma_5) u_i$$

This is why neutral current data have not told us whether neutrinos are Majorana or Dirac![29]

For a relativistic, left-handed neutrino ν_-, the matrix element of the neutrino neutral current in the Dirac case is indistinguishable from its counterpart in the Majorana case.

This example, like the decay $K^+ \to \pi^+ \nu\bar{\nu}$ discussed earlier, illustrates what we call the *Practical Dirac – Majorana Confusion Theorem*:[11]

Suppose the mass of some neutrino is very small compared to all other mass or energy scales in the problem, and that all weak currents are left-handed. Suppose further (as is the case in practice) that all experiments on this neutrino are done with one of the two states: "ν_-" or its CPT conjugate "ν_+". Then it is practically impossible to tell experimentally whether "ν_-" and "$\bar{\nu}_+$" are really (ν_-^M, ν_+^M), the two helicity states of a Majorana neutrino, or $(\nu_-^D, \bar{\nu}_+^D)$, two of the four states of a Dirac neutrino.

When $M_\nu = 0$, there is no distinction at all between $(\nu_-^D, \bar{\nu}_+^D)$ and (ν_-^M, ν_+^M). Indeed, one can no longer flip the helicity of a neutrino by going into another Lorentz frame (nor, as one can show, by weak interactions or an external EM field). So the pair $(\nu_-^D, \bar{\nu}_+^D)$ gets disconnected from the pair $(\bar{\nu}_-^D, \nu_+^D)$, and the latter pair need not even exist. We just have two states, and it makes no difference what we call them. The distinction between Dirac and Majorana becomes purely arbitrary.

Unfortunately, the $M_\nu = 0$ limit is approached smoothly, so that even if $M_\nu \neq 0$, if it is small the distinction between the Dirac and Majorana cases is very difficult to observe.

The conclusion of this section is that the neutral current scattering will not tell if the neutrino is a Majorana or Dirac particle. In fact, the only feasible experiment up to now which can help in solving this problem is the neutrinoless double β decay.

2 - NEUTRINOLESS DOUBLE β DECAY

Let us consider the reaction:

$$(A, Z) \rightarrow (A, Z + 2) + 2\,e^-$$

where a nucleus decays into another nucleus with the emission of two electrons. This process can arise from the following diagram:

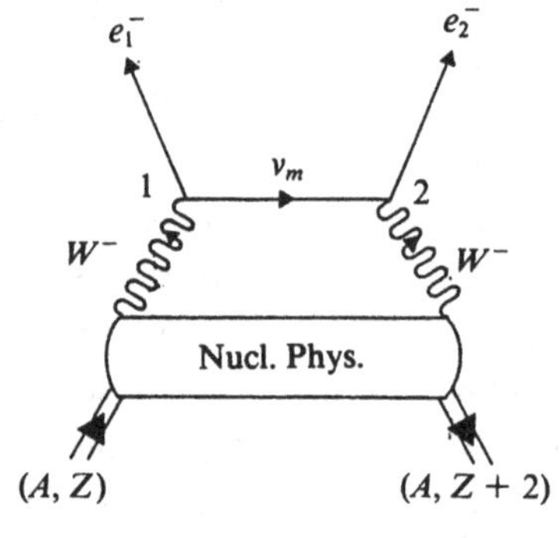

The emission of two W bosons by the initial nucleus is described by some nuclear matrix element. The subsequent production of the electron pair via exchange of a neutrino mass eigenstate ν_m is given by particle physics. In this process, the lepton number is not conserved.

Let us first assume that the charged weak currents are the usual ones, which conserve lepton number and are left-handed. The neutrino emitted at the first vertex looks like a $\bar{\nu}$ while the neutrino absorbed at the second vertex looks like a ν. Therefore, the diagram will not occur unless the intermediate neutrino is a Majorana neutrino, so that $\bar{\nu}$ and ν are the same particle. Now, even for a Majorana neutrino the process is suppressed:

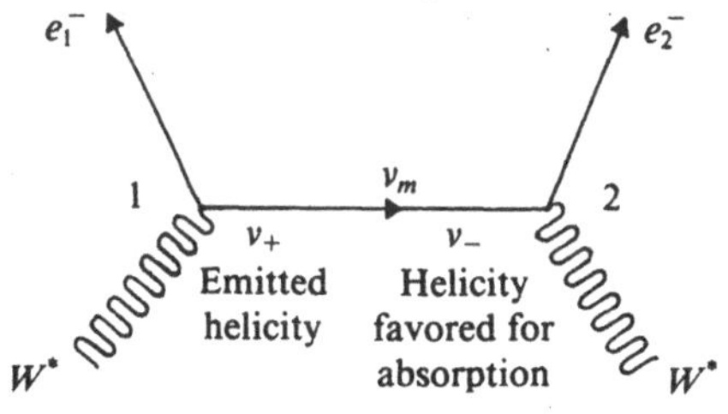

As we discussed earlier, the helicity of the Majorana ν_m will behave as if ν_m were a Dirac particle. Thus, at the first vertex, the neutrino emitted together with an electron is dominantly right-handed, since its helicity behaves like that of a $\bar{\nu}^D$. However, at the second vertex, the same neutrino interacts appreciably only if it is left-handed, since here the current treats it like a ν^D. The amplitude for this neutrino to be emitted *left-handed* at vertex 1, so that it can then be absorbed without suppression at vertex 2, is of order M_m/E_m. The amplitude for the reaction is then proportional to the mass M_m, and vanishes if neutrinos are massless.

We shall discuss later what happens if right-handed currents are present in the weak interaction.

a) Calculation of the diagram assuming no right-handed current

Let us now compute the amplitude of neutrinoless double β decay, which is an illustration of how you would calculate a Feynman diagram with Majorana particles.

We are interested in the amplitude $W^* W^* \to e^- e^-$ described by the following diagram:

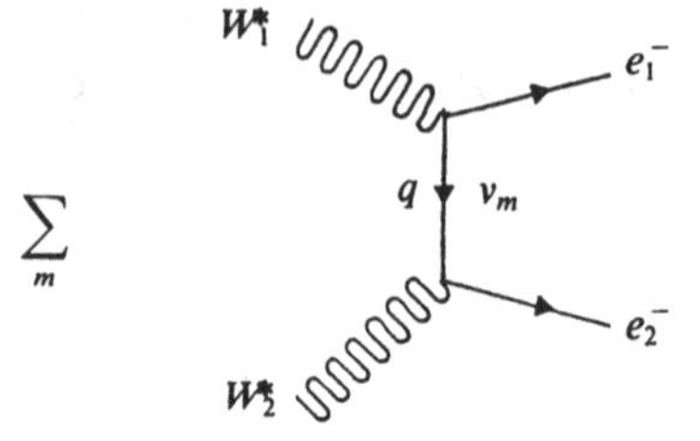

where ν_m is a certain Majorana mass eigenstate, and we are now summing over the contributions of all the ν_m that may exist.

The hamiltonian density of the charged current interaction is (see Chapter II-1):

$$H_{cc} = g\, W_\mu^- \sum_m i\, \bar{e}\, \gamma_\mu(1 + \gamma_5)\, U_{em} \nu_m + g\, W_\mu^+ \sum_m i\, \bar{\nu}_m \gamma_\mu(1 + \gamma_5)\, U_{em}^* e$$

The amplitude contributed by a given ν_m will therefore arise from the second-order term:

$$[W^\mu\, \bar{e}\, \gamma_\mu(1 + \gamma_5)\, U_{em} \nu_m]\, [W^\mu\, \bar{e}\, \gamma_\mu(1 + \gamma_5)\, U_{em} \nu_m] \tag{4.5}$$

We would like to contract the neutrino fields in this expression to form the usual neutrino propagator:

$$\underline{\nu_m \bar{\nu}_m} = \langle 0 |\, T[\nu_m \bar{\nu}_m]\, | 0 \rangle$$

The trouble is that we have no $\bar{\nu}_m$. We could try to find the expression for the unusual propagator obtained by contracting a Majorana field ν_m with itself (rather than with a $\bar{\nu}_m$), but it is more convenient to use a trick which will enable us to work with the normal fermion propagator:

$$\underline{\nu_m \bar{\nu}_m}$$

The trick is to invoke the identity:

$$\begin{aligned}\bar{e}\,\gamma_\mu(1+\gamma_5)\,\nu_m &= \overline{(e^c)^c}\,\gamma_\mu(1+\gamma_5)\,(\nu_m^c)^c\\ &= -(e^c)^t\,\Omega^{-1}\,\gamma_\mu(1+\gamma_5)\,\Omega\left(\overline{\nu_m^c}^{\,t}\right)\\ &= -(e^c)^t\left[-\gamma_\mu(1-\gamma_5)\right]^t\left(\overline{\nu_m^c}^{\,t}\right)\\ &= -\,\overline{\nu_m^c}\,\gamma_\mu(1-\gamma_5)e^c\end{aligned}$$

The minus sign in the last line comes from the anticommutation of two fermion fields.

Now, ν_m is a Majorana field. Thus, from Eq. (3.29), and the relation $C\,\Psi\,C^{-1} = \eta_C^*\,\Psi^c$, we have:

$$\nu_m^c = \lambda_m^*\,\nu_m$$

where λ_m is the creation phase factor of the field ν_m. Hence we see that:

$$\bar{e}\,\gamma_\mu(1+\gamma_5)\nu_m = -\lambda_m\bar{\nu}_m\,\gamma_\mu(1-\gamma_5)e^c$$

Using this relation to rewrite the second factor in expression (4.5), we obtain a second-order interaction which does involve a ν_m and a $\bar{\nu}_m$, so we may now contract them as usual, obtaining:

$$-\left[W^\mu\,\bar{e}\,\gamma_\mu(1+\gamma_5)\,U_{em}\underbrace{\nu_m\right]\left[\lambda_m\,W^\nu\,\bar{\nu}_m}\gamma_\nu(1-\gamma_5)\,e^c\,U_{em}\right]$$

It is easy to verify that $\underbrace{\nu_m\,\bar{\nu}_m}$ gives precisely the usual fermion propagator,

$$\frac{-i\not{q} + M_m}{q^2 + M_m^2}$$

even though ν_m is a Majorana field.

The matrix element between the initial and final states of the interaction above picks up the spinors associated with the creation operators of the electrons. Recall that the electron fields have the structure:

$$e \sim \sum [fu + \bar{f}^{\dagger} v]$$

$$\bar{e} \sim \sum [f^{\dagger} \bar{u} + \bar{f} \bar{v}]$$

and

$$e^{c} \sim \sum [\bar{f} u + f^{\dagger} v]$$

Thus, treating the W bosons as external particles described by polarization vectors ε_i^{μ}, and omitting irrelevant constants, the matrix element becomes

$$\varepsilon_1^{\mu}\, \bar{u}_1 \gamma_{\mu}(1 + \gamma_5)\, \frac{(-i\not{q} + M_m)}{q^2 + M_m^2}\, \gamma_{\nu}(1 - \gamma_5)\, v_2\, \varepsilon_2^{\nu}\, \lambda_m\, U_{em}^2 \tag{4.6}$$

Amusingly, one electron is here being described by a v spinor, rather than by the usual u!

In order to combine the $(1 \pm \gamma_5)$ factors, we use:

$$(1 + \gamma_5)\,(-i\not{q} + M_m)\,\gamma_{\nu} = M_m \gamma_{\nu}(1 - \gamma_5) - i\not{q}\,\gamma_{\nu}(1 + \gamma_5)$$

But

$$(1 - \gamma_5)\,(1 + \gamma_5) = 0$$

$$(1 - \gamma_5)^2 = 2(1 - \gamma_5)$$

Hence, neglecting M_m^2 compared to q^2 in the denominator of the propagator, the amplitude is:

$$2\, \bar{u}_1\, \not{\varepsilon}_1\, \frac{M_m}{q^2}\, \not{\varepsilon}_2 (1 - \gamma_5)\, v_2\, \lambda_m\, U_{em}^2$$

We now have to add the contribution of the graph where electrons 1 and 2 are exchanged and to sum over all Majorana species ν_m. The resulting amplitude for neutrinoless double β decay has the form:

$$A\,[\beta\beta_{0\nu}] = \left[\sum_m \lambda_m M_m U_{em}^2\right] \times [m\text{-Independent Factor}] \tag{4.7}$$

where the second factor includes the highly nontrivial nuclear physics of the process.

We have learned here that when dealing with Majorana fermions, a good way to proceed is to forget about the usual Feynman rules and to construct the amplitudes from the field operators.[30]

Looking at formula (4.7), we might be surprised that the arbitrary phase factor λ_m appears in the decay rate. This in fact makes sense because the matrix element U_{em} also depends on this arbitrary phase factor, and it is $\lambda_m U_{em}^2$ which does not. Let us explain this in more detail.

Let us assume that CP is conserved. Under CP, any fermion field transforms according to:

$$CP\,\Psi(\mathbf{x}, t)\,(\mathrm{CP})^{-1} = \eta_{\mathrm{CP}}^{*}\,\gamma_4\,\gamma_2\,\Psi^{*}(-\mathbf{x}, t) \tag{4.8}$$

where η_{CP} is a phase factor. One can show (see exercise III.5) that this transformation law for the field implies the following transformation law for the corresponding state $|f\rangle$, depending on whether Ψ is a Dirac or a Majorana field:

For Ψ a Dirac field: $\quad \mathrm{CP}\,|f(\mathbf{p}, s)\rangle = \eta_{\mathrm{CP}}\,|\bar{f}(-\mathbf{p}, s)\rangle \qquad$ (4.9a)

For Ψ a Majorana field: $\quad \mathrm{CP}\,|f(\mathbf{p}, s)\rangle = \tilde{\eta}_{\mathrm{CP}}\,|f(-\mathbf{p}, s)\rangle$

$$\text{with } \tilde{\eta}_{\mathrm{CP}} = \eta_{\mathrm{CP}}\lambda = -\tilde{\eta}_{\mathrm{CP}}{}^{*} \tag{4.9b}$$

In Eq. (4.9b), λ is the creation phase factor for the Majorana field in question. (Note that $\tilde{\eta}_{\mathrm{CP}}$, the intrinsic CP parity of a Majorana fermion, is indeed imaginary, as we previously proved it must be, at least when the fermion is a neutrino, by considering $Z^{\circ} \rightarrow \nu^M \nu^M$.)

With the obvious behavior of W_μ^- under CP, Eq. (4.8) implies that the CP transform of the first term in our charged current Hamiltonian density, written just above Eq. (4.5), is:

$$\mathrm{CP}\sum_m \left[W_\mu^-\, i\bar{e}\,\gamma_\mu(1+\gamma_5)\,U_{em}\,\nu_m\right]_{\mathbf{x},t}(\mathrm{CP})^{-1}$$

$$= \sum_m \left[\eta_{\mathrm{CP}}(W)\,\eta_{\mathrm{CP}}(e)\,\eta_{\mathrm{CP}}^{*}(\nu_m)\right]\left[W_\mu^+\, i\bar{\nu}_m\,\gamma_\mu(1+\gamma_5)\,U_{em}\,e\right]_{-\mathbf{x},t}$$

That is, apart from some constants and $\mathbf{x} \to -\mathbf{x}$, the two terms in H_{CC} go into each other under CP. Now, the change $\mathbf{x} \to -\mathbf{x}$ is irrelevant since one will eventually integrate over $\mathbf{x}$. Thus, CP invariance of H_{CC} requires only that the constants match up; that is, that

$$U_{em}^* = U_{em}\left[\eta_{CP}(W)\,\eta_{CP}(e)\,\eta_{CP}^*(\nu_m)\right] \tag{4.10}$$

i) In the case of Dirac neutrinos, every phase factor η_{CP} in Eq. (4.10) can be absorbed into the definition of the antiparticle state $|\bar{f}\rangle$ relative to the particle state $|f\rangle$. Then the U matrix is real.

ii) In the case of Majorana neutrinos, let us choose

$$\lambda_m = \frac{\tilde{\eta}_{CP}(\nu_m)}{i} \tag{4.11}$$

where $\tilde{\eta}_{CP}(\nu_m)$ is the intrinsic CP parity of the ν_m state.

Then, from Eqs. (4.10) and (4.9b),

$$\frac{U_{em}^*}{U_{em}} \propto \left[\frac{\tilde{\eta}_{CP}(\nu_m)}{\lambda_m}\right]^* \text{ is independent of } m!$$

We can then choose $\eta_{CP}(W)\,\eta_{CP}(e)$ so that

$$U_{em}^* = U_{em} \tag{4.12}$$

In summary, *if CP is conserved, the mixing matrix U can be chosen real.* However, in the Majorana case, this choice involves choosing nontrivial values for the creation phase factors λ_m in the Majorana neutrino fields. These nontrivial λ_m will then appear in reaction amplitudes (as, for example, in Eq. (4.7)).

Now, for any choice of the λ_m, Eqs. (4.10) and (4.9b) imply that:

$$\begin{aligned}\lambda_m U_{em}^2 &\propto \lambda_m U_{em} U_{em}^* \eta_{CP}(\nu_m) \\ &= \tilde{\eta}_{CP}(\nu_m)\,|U_{em}|^2\end{aligned}$$

where $\tilde{\eta}_{CP}(\nu_m)$ is the phase-convention independent intrinsic CP parity of the Majorana neutrino ν_m. The amplitude (4.7) of the double β decay has, therefore, no dependence on arbitrary phases:[31]

$$A\,[\beta\beta_{0\nu}] \propto \sum_m \underbrace{\frac{\tilde{\eta}_{CP}(\nu_m)}{i}}_{\text{real number}} |\,U_{em}\,|^2 M_m \tag{4.13}$$

Assuming that the nuclear factor in Eq. (4.7) is known, a measurement of a double β decay amplitude will determine the quantity on the right side of Eq. (4.13), which we refer to as the effective neutrino mass M_{eff}:

$$M_{\text{eff}} = \sum_m \frac{\tilde{\eta}_{CP}(\nu_m)}{i} |\,U_{em}\,|^2 M_m \tag{4.14}$$

Owing to the CP parities in M_{eff}, this effective mass can be smaller than all the masses M_m,

$$M_{\text{eff}} << M_m$$

even though U is unitary. Indeed, the cancellations in M_{eff} which can make this quantity anomalously small do occur in some popular grand unified theories.[32]

In summary, even if neutrinos are massive Majorana particles, the amplitude for neutrinoless double β decay can be very small compared to the neutrino masses. The failure to observe this decay at a given level does not imply an upper bound on neutrino mass.

(The experimental limit on M_{eff} is already quite small:[33]

$$M_{\text{eff}} < (1 - 2)\ \text{eV}$$

Due to the possible cancellations, even if neutrinos are of Majorana character this limit does not necessarily contradict the evidence from the ITEP experiment[5] that the dominant mass eigenstate in ν_e has a mass between 17 and 40 eV. However, the cancellation which is needed to avoid the contradiction appears to be quite unlikely in view of existing laboratory data and cosmological constraints.[34] In addition, the SIN upper limit of 18 eV[35] on the mass of the dominant mass eigenstate in ν_e is in rather strong disagreement with the ITEP evidence. It will be interesting to see what future experimental results will show.)

b) Neutrinoless double β decay with right-handed weak charged currents

Let us suppose that there are other W bosons than the one discovered at CERN. In general, each boson W_a may couple to leptons via left-handed (L), right-handed (R), or both kinds of currents. For example, in the left-right symmetric models, there is a second W boson with right-handed coupling to the fermions. When there are several W bosons, the neutrino exchange diagram for double β decay becomes:

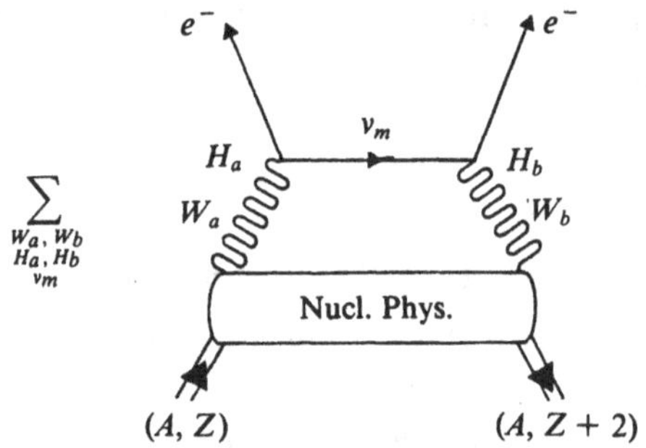

Here H_a is the handedness of the current to which W_a couples. We must now sum, not only over all possible neutrinos, but also over all possible W_a, W_b, and, for each W_a and W_b, over all possible handednesses H_a, H_b of the currents to which they couple.

As before, the intermediate neutrino has to be a Majorana neutrino. What is really new here is that there is, for example, a nonvanishing amplitude for a left-handed coupling at vertex (a) and a right-handed coupling at vertex (b). One might guess that, with this combination of couplings, one can get around the helicity suppression we talked about before: a right-handed neutrino can be emitted at vertex (a) and be absorbed at vertex (b), without suppression, and seemingly without any need for the neutrino to have a mass. The conclusion would then be: if there is a right-handed weak current, double β decay can occur even if all neutrinos are massless. This, however, is not true so long as the weak interactions are described by a gauge theory.[36]

In order to explain that, let us first consider the scattering $\nu\bar{\nu} \to W^+ W^-$ described by graph 1 below. Here W is the ordinary weak boson, which couples to left-handed currents, as shown. Now, graph 1 has very bad high-energy behavior: when the energy goes to infinity, the graph leads to a cross section which violates the unitarity bound. However, it is a hallmark of gauge theories that the *complete* lowest-order amplitude for any process like $\nu\bar{\nu} \to W^+ W^-$ has no such unacceptable behavior. In the standard model of the weak interactions, the acceptable high-energy behavior results from the presence of a second diagram: the $Z°$ pole (graph 2). This graph precisely cancels the bad high-energy behavior from the electron exchange of graph 1.

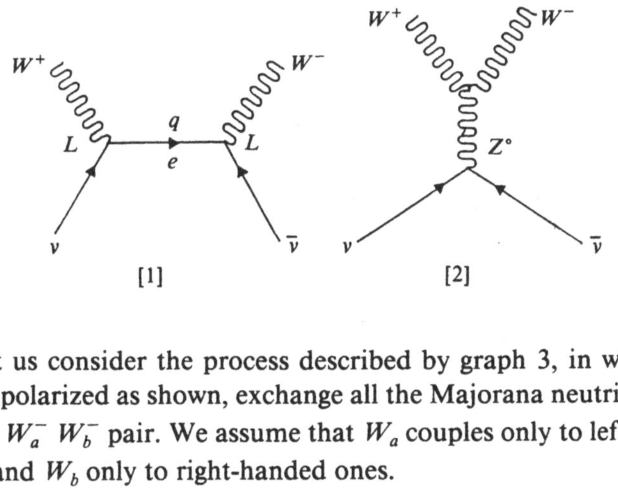

[1] [2]

Now let us consider the process described by graph 3, in which two electrons, polarized as shown, exchange all the Majorana neutrinos ν_m to produce a $W_a^- W_b^-$ pair. We assume that W_a couples only to left-handed currents, and W_b only to right-handed ones.

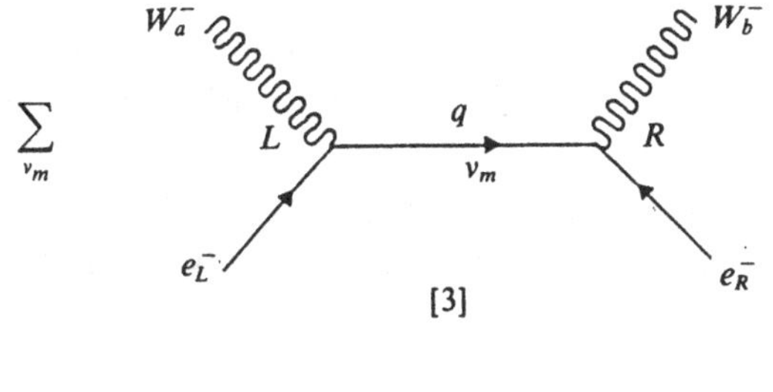

[3]

Apart from a time reversal, graph 3 is just the $H_a H_b = LR$ term in double β decay. We have said that this term does not seem to require neutrino mass. Indeed, if we calculate the amplitude for graph 3, we find that any one ν_m exchange does not vanish when the mass $M_m \rightarrow 0$. However, in a gauge theory, if all $M_m = 0$, *the various* ν_m *exchanges will precisely cancel each other when we add them up.*[36] To see why they will do so, we note that, as one can show, any one ν_m exchange in graph 3 has exactly the same unitarity-violating high-energy behavior as the electron exchange in graph 1. Now, in any gauge theory, the *complete* lowest-order amplitude for $e_L^- e_R^- \rightarrow W_a^- W_b^-$ will not violate unitarity. However, in $e_L^- e_R^- \rightarrow W_a^- W_b^-$, no $Z°$ pole like that in graph 2 can contribute, because the initial state is doubly charged. Thus, assuming there are no double-charged Z bosons, the unitarity-respecting high-energy behavior of $e_L^- e_R^- \rightarrow W_a^- W_b^-$ can only result from a cancellation among the various ν_m exchanges themselves.

The amplitude corresponding to graph 3 has the form:

$$\sum_{\nu_m} U_{em}^{(a)} \frac{\not{q}}{q^2 + M_m^2} U_{em}^{(b)} \tag{4.15}$$

Here $U^{(a)}$ is the analogue for W_a of the mixing matrix U for the ordinary W (83 GeV), and $\not{q}/(q^2 + M_m^2)$ is the contributing part of the ν_m propagator. From (4.15), we see that the required cancellation among the ν_m exchanges when the energy and q^2 are large implies that:

$$\sum_m U_{em}^{(a)} U_{em}^{(b)} = 0 \tag{4.16}$$

But then, if all the neutrino masses are zero, the amplitude (4.15) vanishes at all energies. Therefore, any contribution to the double β decay amplitude with a right-handed coupling at one of the lepton vertices and a left-handed one at the other vanishes, unless some neutrino has a nonzero mass.[37]

The conclusion is then: if (1) neutrinoless double β decay results from neutrino exchange, and (2) there are no doubly-charged gauge bosons, then the observation of this decay would imply the existence of a massive Majorana neutrino, whether there are right-handed currents or not. Indeed, it can be shown that observation of neutrinoless double β decay would imply a massive neutrino even if conditions (1) and (2) do not hold, but then the neutrino may be exceedingly light.[38] On the other hand, from the argument we have given above, which does entail the assumptions (1) and (2), one can go on to show that observation of neutrinoless double β decay would actually imply a significant lower bound on the mass of the heaviest neutrino.[36] To do that, one must make the additional (highly-plausible) assumption that the effective strength, G_{Right}, of any right-handed weak interaction is not greater than the effective strength, G_{Fermi}, of the known left-handed weak interaction. The lower bound which then follows can be expressed in terms of the lifetime for the double β decay. If, for example, ^{76}Ge were found to undergo neutrinoless double β decay with a lifetime τ_{Ge}, that would imply that the heaviest neutrino has a mass M satisfying:

$$M \gtrsim 1 \text{ eV} \left[10^{24}\,\text{yr}/\tau_{\text{Ge}}\right]^{1/2}$$

The current limit on τ_{Ge}, which is our most stringent limit on the particle physics of neutrinoless double β decay, is $\tau_{\text{Ge}} > 5 \times 10^{23}$ yr.[39] Thus, if neutrinoless double β decay of ^{76}Ge were to be seen at a rate not far below the current bound, it would follow that at least one neutrino has $M \gtrsim 1$ eV. Such a mass is large enough to be sought in neutrino oscillation experiments, and possibly also in the next generation of tritium β decay experiments.

In conclusion, if neutrinoless double β decay is *not* seen at a certain level, its absence does *not* imply an upper bound on neutrino mass, but if it *is* seen, its presence *does* imply a lower bound on neutrino mass.

EXERCISES

1 – Suppose there exists a W boson, W_R, which is much heavier than the known W boson, W (83 GeV). Suppose that W_R couples with normal weak strength to right-handed quark currents, and to a right-handed lepton current which couples the electron to a heavy neutral Majorana

particle N rather than to the neutrino. Assume that $M_N >> 10$ MeV, the typical momentum transfer in $\beta\beta_{o\nu}$. Consider, now, the amplitude A_N for $\beta\beta_{o\nu}$ contributed by a diagram like the first one in Chapter IV.2, but with each W(83 GeV) replaced by a W_R, and the exchanged neutrino replaced by an N. Assuming that $M_{W_R} = 1.6$ TeV (see discussion in Chapter V.2), use the fact that the M_{eff} defined by Eq. (4.14) is experimentally less than ≈ 1 eV to set a *very crude* lower bound on the mass of N. (Neglect possible differences between the nuclear matrix element in A_N and that in the amplitude corresponding to the first diagram in Chapter IV.2.)[40]

2 – Assume that the $Z°$ couples to any neutrino with the standard-model coupling:

$$H_{\text{NC}} = Z^{\circ}_{\mu}\,\overline{\nu}\,\gamma_{\mu}\,(1 + \gamma_5)\,\nu$$

Imagine that there is a particular neutrino ν whose mass M may be as large as $M_Z/2$, and consider the decay of a $Z°$ with polarization vector ε_μ into a $\nu\nu$ pair. What is the amplitude A_M, analogous to that of Eq. (4.3), for this decay if ν is a Majorana particle? By contrast, what is the amplitude A_D for the decay if ν is a Dirac particle (so that the decay is really into $\nu\overline{\nu}$)? Without detailed calculations, show that when the outgoing neutrinos in the decay are highly relativistic, A_M and A_D lead to identical decay rates in the two cases.

Now suppose M is nearly $M_Z/2$, so that the outgoing neutrinos in the decay are nonrelativistic. We know that if ν is a Majorana particle, the final $\nu\nu$ state will be $^3P_1$, since that is the only antisymmetric $J = 1$ state. Considering that the amplitude for decay into a state with orbital angular momentum L is proportional to p^L (where p is the outgoing momentum in the $Z°$ rest frame), what state does one expect to be the dominant final state when ν is a Dirac particle? Now obtain the nonrelativistic limits of A_M and A_D, using Eqs. (3.11), (3.13) and (3.15). From these limits and Eq. (3.12), obtain the ratio Γ_M/Γ_D of decay rates in the Majorana and Dirac cases when the outgoing neutrinos are nonrelativistic.

CHAPTER

FIVE

NEUTRINO MASSES IN GAUGE THEORIES

We said at the beginning of the lectures that neutrinos are likely massive because this is what the grand unified models lead us to expect. You are probably very eager to learn what these gauge theories predict for *the values* of the neutrino masses. Unfortunately, gauge theories do not have any precise predictions for these values. Nor, for that matter, do they have any precise predictions for any fermion masses.

It is fair to say that gauge theories accommodate, rather than predict, fermion masses. However, they accommodate these masses in ways that constrain them or relate them to other things. Some of these theories "explain" why neutrino masses are so small by relating them, via the "see-saw relation" (1.1), to the inverse of some large mass scale.[1] The see-saw relation occurs, for example, in SO(10) grand unified models, and in left-right symmetric gauge theories. Now, if the large mass scale M_N in Eq. (1.1) is the GUT scale, 10^{14} GeV, and the typical quark or charged lepton mass $M_{q \text{ or } \ell}$ is taken to be 1 GeV, then, as we said before, $M_\nu = 10^{-5}$ eV. Thus, the see-saw relation (1.1) not only provides an understanding of why neutrinos are so light, but suggests that they may be much too light for their masses to be detected by present laboratory methods. On the other hand, we don't know much about M_N and, in practice, neutrino masses can be anything. In left-right symmetric models, one often considers M_N values in the (10–100) GeV range. Such values lead, of course, to neutrino masses much larger than 10^{-5} eV in Eq. (1.1). In addition, since M_N is the mass of a predicted heavy neutral lepton (see discussion before Eq. (1.1)), these M_N values imply that we may be able to detect this heavy neutral lepton at accelerators of the not-too-distant future.

In this last chapter, we shall study the basic left-right symmetric model as an illustration of the treatment of neutrino masses in gauge theories. We shall, in particular, derive the see-saw relation among fermion masses.

Our discussion will often be in terms of chirally left- or right-handed fields $\psi_{L,R}$ which satisfy:

$$P_{L,R}\,\psi_{L,R} \equiv \frac{1 \pm \gamma_5}{2}\,\psi_{L,R} = \psi_{L,R} \quad ; \quad P_{R,L}\,\psi_{L,R} = 0$$

In a gauge theory, such chiral fields are put into multiplets of the group on which the theory is based.

The left-right symmetric theory[41] is based on the group $SU(2)_L \times SU(2)_R \times U(1)$. The $SU(2)_L$ is referred to as the left-handed isospin (I_L) group, and the $SU(2)_R$ as the right-handed isospin (I_R) group. This model contains the standard weak interaction coupling the W(83 GeV) to the left-handed quarks and leptons, which are in I_L doublets:

$$\begin{bmatrix} u_L \\ d_L \end{bmatrix} , \begin{bmatrix} \psi_L \\ e_L \end{bmatrix}$$

Here ψ_L is the left-handed ν_e field (we shall discuss only one generation).

As in the standard model, the $I_L = 1$ boson W (83 GeV) does not couple to the right-handed quarks and leptons, which are I_L singlets. However, in addition to the W(83 GeV), which we shall call W_L, the left-right symmetric model contains a second, heavier W, W_R. The W_R, an $I_R = 1$ analogue of the W_L, does couple to the right-handed fermions, which are placed in doublets of I_R:

$$\begin{bmatrix} u_R \\ d_R \end{bmatrix} , \begin{bmatrix} \psi_R \\ e_R \end{bmatrix}$$

Here ψ_R is a right-handed neutrino field. The coupling of the W_R to the right-handed doublets is exactly the same as that of W_L to the left-handed ones, in conformity with the principle of left-right symmetry.

Now, one *could* assume that ν_e is a Dirac particle, so that, after the fermions acquire mass, $\psi_L + \psi_R \equiv \psi$ is simply the Dirac field for a ν_e of a certain mass, just as $e_L + e_R \equiv e$ is the Dirac field for the electron. However, this assumption, which would treat the ν_e in the same way as the e, u, and d are treated, would not explain why the ν_e is so much lighter than all these other fermions. Therefore, one treats the neutrino in a special way, which we shall now discuss.

1 - NEUTRINO MASS TERMS IN FIELD THEORY

Suppose ψ is some *Dirac* field, and $\psi_{L,R} \equiv [(1 \pm \gamma_5)/2]\psi$ are its left- and right-handed projections. In general, the Lagrangian density for the theory containing ψ will include a mass term of the form:

$$\begin{aligned} -L_m &= M\,\bar{\psi}\,\psi \\ &= M\,(\bar{\psi}_R\,\psi_L + \bar{\psi}_L\,\psi_R) \\ &= M\,(\bar{\psi}_R\,\psi_L + h.c.) \end{aligned} \qquad (5.1)$$

where *h.c.* stands for the hermitian conjugate of the previous term.

This is called a *Dirac mass term*. It leads to the usual mass term in the Dirac equation for ψ.

Now, can we find additional, equally Lorentz-invariant expressions which are similar to the mass term (5.1), and are potential additional mass terms, by multiplying one of the fields

$$\psi_L\,,\ \psi_R\,,\ (\psi^c)_L\,,\ (\psi^c)_R$$

by one of the fields

$$\bar{\psi}_L\,,\ \bar{\psi}_R\,,\ \overline{(\psi^c)}_L\,,\ \overline{(\psi^c)}_R\ ?$$

We have to keep in mind that the two fields we choose must have opposite chiralities, since a product like $\bar{\chi}_L \psi_L$ vanishes:

$$\bar{\chi}_L \psi_L = \chi_L^\dagger \gamma_4 \psi_L = \chi^\dagger \frac{1+\gamma_5}{2} \gamma_4 \psi_L = \chi^\dagger \gamma_4 \frac{1-\gamma_5}{2} \psi_L = 0$$

(γ_5 is hermitian and anticommutes with γ_4.) This rule is the reason the combinations $\bar{\psi}_L \psi_L$ and $\bar{\psi}_R \psi_R$ do not appear in Eq. (5.1).

Using only ψ_L and ψ_R and their adjoints, we get once more $\bar{\psi}_L \psi_R$ and $\bar{\psi}_R \psi_L$, which is the hermitian conjugate of $\bar{\psi}_L \psi_R$.

Using only $(\psi^c)_L$ and $(\psi^c)_R$ and their adjoints we get:

$$\overline{(\psi^c)_L}\, (\psi^c)_R$$

and its hermitian conjugate.

Let us now establish a useful property:

$$\begin{aligned}
(\psi_L)^c &= \Omega\, \bar{\psi}_L^t \\
&= \Omega\, \gamma_4 \left[\frac{1+\gamma_5}{2} \psi \right]^* \\
&= \Omega\, \gamma_4 \frac{1+\gamma_5}{2} \psi^* \qquad (\gamma_5 \text{ is real}) \\
&= \frac{1-\gamma_5}{2} \Omega\, \gamma_4\, \psi^* \\
&= (\psi^c)_R
\end{aligned}$$

(γ_5 commutes with Ω and anticommutes with γ_4.)

Namely:

$$(\psi_{L,R})^c = (\psi^c)_{R,L} \tag{5.2}$$

Then we have:

$$\overline{(\psi^c)}_L\, (\psi^c)_R = \overline{(\psi_R)^c}\, (\psi_L)^c$$

$$= -\psi_R^t \, \Omega^{-1} \, \Omega \, \bar{\psi}_L^t \qquad \text{(see Table 3)}$$

$$= -\psi_R^t \, \bar{\psi}_L^t$$

$$= \bar{\psi}_L \, \psi_R \tag{5.3}$$

because fermion fields anticommute.

Conclusion: with $(\psi^c)_L$ and $(\psi^c)_R$ and their adjoints we get again $\bar{\psi}_L \psi_R$ and $\bar{\psi}_R \psi_L$. Thus, so far we have found nothing new.

However, there remain the combinations:

$$\overline{\psi^c}_L \psi_R \,, \; \overline{\psi^c}_R \psi_L \,, \; \bar{\psi}_L (\psi^c)_R \,, \quad \text{and} \quad \bar{\psi}_R (\psi^c)_L \tag{5.4}$$

Now, for any two spinor fields χ and φ, $(\bar{\chi}\varphi)^\dagger = \bar{\varphi}\chi$, so the second two combinations are just the hermitian conjugates of the first two. Thus, we conclude that the distinct combinations which are available for building mass terms in the Lagrangian are:

$$\bar{\psi}_L \psi_R \,, \; \overline{\psi^c}_R \psi_L \,, \; \overline{\psi^c}_L \psi_R$$

and their hermitean conjugates. Introducing three constants M_D, M_L, and M_R with the dimensions of mass, we may write the most general sum of mass terms as

$$\begin{aligned} -L_m &= M_D \left[\bar{\psi}_L \psi_R + h.c. \right] \\ &\quad + \frac{M_L}{2} \left[(\overline{\psi^c})_R \psi_L + h.c. \right] \\ &\quad + \frac{M_R}{2} \left[(\overline{\psi^c})_L \psi_R + h.c. \right] \end{aligned} \tag{5.5}$$

Note that the Dirac mass term is not the only possible mass term. We have two new terms, which are called *Majorana mass terms.*

In writing Eq. (5.5), we have taken all three mass parameters to be real. It can be shown that we may always do so if we are content with a CP invariant theory.

Adding to Eq. (5.5) the usual kinetic energy term, we can now write the general free-field Lagrangian for the field ψ as

$$-L = \bar{\psi}\gamma_\mu \partial_\mu \psi + M_D[\bar{\psi}_L \psi_R + h.c.] + \frac{M_L}{2}\left[\overline{(\psi_L)^c}\psi_L + h.c.\right] + \frac{M_R}{2}\left[\overline{(\psi_R)^c}\psi_R + h.c.\right] \tag{5.6}$$

It will be easier to understand the physics content of this Lagrangian if we rewrite it in terms of two new fields f and F:[42]

$$f = \frac{\psi_L + (\psi_L)^c}{\sqrt{2}} \qquad F = \frac{\psi_R + (\psi_R)^c}{\sqrt{2}} \tag{5.7}$$

Then we have:

$$\bar{f}f = \frac{1}{2}\left[\bar{\psi}_L\psi_L + \overline{(\psi_L)^c}\psi_L + \bar{\psi}_L(\psi_L)^c + \overline{(\psi_L)^c}(\psi_L)^c\right]$$

$$= \frac{1}{2}\left[\overline{(\psi_L)^c}\psi_L + h.c.\right]$$

$$\bar{f}F = \frac{1}{2}\left[\bar{\psi}_L\psi_R + \overline{(\psi_L)^c}\psi_R + \bar{\psi}_L(\psi_R)^c + \overline{(\psi_L)^c}(\psi_R)^c\right]$$

$$= \frac{1}{2}\left[\bar{\psi}_L\psi_R + \bar{\psi}_R\psi_L\right]$$

and $$\bar{F}f = \frac{1}{2}\left[\bar{\psi}_L\psi_R + \bar{\psi}_R\psi_L\right]$$

In terms of the new fields, the Lagrangian becomes:

$$-L = \bar{f}\gamma_\mu\partial_\mu f + \bar{F}\gamma_\mu\partial_\mu F + M_D(\bar{f}F + \bar{F}f) + M_L\bar{f}f + M_R\bar{F}F$$

$$= \bar{f}\gamma_\mu\partial_\mu f + \bar{F}\gamma_\mu\partial_\mu F + (\bar{f}, \bar{F})\begin{bmatrix} M_L & M_D \\ M_D & M_R \end{bmatrix}\begin{bmatrix} f \\ F \end{bmatrix}$$

In rewriting the kinetic energy term, we have used the fact that for any two fields χ and φ, $\bar{\chi}_{L,R}\gamma_\mu\partial_\mu\varphi_{R,L} = 0$, and carried out a certain amount of algebra.

The Lagrangian is now of the form:

$$-L = \bar{V}\not{\partial}V + \bar{V}[M]V \tag{5.8}$$

where V is the column vector

$$\begin{bmatrix} f \\ F \end{bmatrix} \tag{5.9}$$

and $[M]$ is the real symmetric matrix

$$[M] = \begin{bmatrix} M_L & M_D \\ M_D & M_R \end{bmatrix}$$

The matrix $[M]$ is called the *neutrino mass matrix*.

This matrix can be diagonalized by a rotation in the two-dimensional space of the vector V.

Let ν' and N be the eigenvector fields, with eigenvalues $M_{\nu'}$ and M_N. Then:

$$-L = \bar{\nu}'\not{\partial}\nu' + \bar{N}\not{\partial}N + M_{\nu'}\bar{\nu}'\nu' + M_N\bar{N}N \tag{5.10}$$

The physics content of this Lagrangian is now clear: it is the free Lagrangian for *two* particles, ν' and N, with masses $M_{\nu'}$ and M_N. Recall that the Dirac field ψ in terms of which we originally wrote the Lagrangian corresponded to 4 states: a Dirac field normally describes the 2 spin states of a particle plus the 2 spin states of its (distinct) antiparticle. Since we now have *two* particles ν' and N with different masses, we expect that each of them must have only 2 spin states. That is, we expect that each of them is a CPT self-conjugate Majorana particle. We shall prove later that this is indeed the case: the inclusion of the Majorana mass terms in Eq. (5.6) has split the four degenerate states of the field ψ into two nondegenerate Majorana pairs.

In a gauge theory, the Majorana and Dirac mass terms come from couplings of the fermions to Higgs fields. Let us discuss this in our illustrative left-right symmetric model.

2 - THE SEE-SAW MECHANISM

Let us first recall that the Lagrangian of a gauge theory contains Yukawa couplings between a fermion field ψ and a scalar Higgs fields Φ of the form:

$$\bar{\psi}\,\Phi\,\psi$$

Such couplings are allowed by gauge invariance.

During the spontaneous symmetry breaking, the field Φ acquires some vacuum expectation value $<\Phi>_0$. The Yukawa coupling then gives:

$$<\Phi>_0\,\bar{\psi}\,\psi$$

which is a mass term, the vacuum expectation $<\Phi>_0$ being proportional to the mass.

In our left-right symmetric model, the spontaneous symmetry breaking is a two step process. First, the gauge group $SU(2)_L \times SU(2)_R \times U(1)$ is broken down to $SU(2)_L \times U(1)$ and the field W_R acquires a mass $M(W_R)$ which is of the order of the symmetry breaking scale $<\Phi>_R$.

$$SU(2)_L \times SU(2)_R \times U(1) \xrightarrow{<\Phi>_R} SU(2)_L \times U(1)$$

yielding $$M_{W_R} \sim <\Phi>_R$$

Then the group $SU(2)_L \times U(1)$ is broken down to the $U(1)$ factor of electromagnetism and the usual W_L field acquires its mass $M(W_L)$, measured around 83 GeV at the CERN collider.

$$SU(2)_L \times U(1) \xrightarrow{<\Phi>_L} U(1)_{EM}$$

yielding $$M_{W_L} \sim 83\ \text{GeV} \sim <\Phi>_L$$

No effects of right-handed currents have been seen experimentally, so we expect the mass of the W_R to be very large. For example, the W_R would contribute to the amplitude $\langle K^0 | H | \bar{K}^0 \rangle$ through the graph of Figure 7, and from the experimental value of the mass difference between the K°_L and the K°_S mesons, one can set the limit:[43]

$$M_{W_R} > 1.6\ \text{TeV}$$

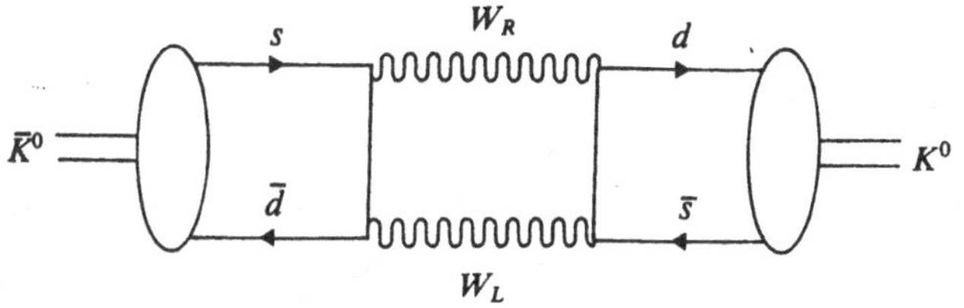

Fig. 7. Contribution of the W_R to the amplitude $\langle K^0 | H | \bar{K}^0 \rangle$.

The breaking scale of $SU(2)_R$ is thus expected to be much bigger than the one of $SU(2)_L$:

$$<\Phi>_R >> <\Phi>_L$$

Let us now construct Yukawa couplings.

The neutrino fields ψ_L and ψ_R in the left-right symmetric model have $SU(2)_L \times SU(2)_R \times U(1)$ quantum numbers. As we have said earlier:

the field ψ_L is in an $SU(2)_L$ doublet $\begin{bmatrix} \psi_L \\ e_L \end{bmatrix}$

with $\quad I_L = \frac{1}{2} \qquad I_R = 0$

the field ψ_R is in an $SU(2)_R$ doublet $\begin{bmatrix}\psi_R\\ e_R\end{bmatrix}$

with $\quad I_L = 0 \quad I_R = \frac{1}{2}$

Therefore, the bilinear products involving ψ_L and ψ_R have the following quantum numbers:

$$\bar{\psi}_R \psi_L \text{ has } I_L = \frac{1}{2} \quad I_R = \frac{1}{2}$$

$$\overline{(\psi_L)^c}\psi_L = \psi_L^t \Omega\, \psi_L \text{ has } I_L = 1 \quad I_R = 0$$

$$\overline{(\psi_R)^c}\psi_R = \psi_R^t \Omega\, \psi_R \text{ has } I_L = 0 \quad I_R = 1$$

These terms, with constant coefficients, are forbidden by the $SU(2)_L \times SU(2)_R \times U(1)$ invariance of the Lagrangian. However, allowed terms (with total $I_L = I_R = 0$) may be constructed by coupling these bilinears to Higgs fields Φ, Δ_L, and Δ_R with properly chosen quantum numbers. These Yukawa couplings are:

$$\bar{\psi}_R\, \Phi\, \psi_L \text{ where } \Phi \text{ has } I_L = 1/2 \quad I_R = 1/2$$

$$\overline{(\psi_L)^c}\, \Delta_L\, \psi_L \text{ where } \Delta_L \text{ has } I_L = 1 \quad I_R = 0$$

$$\overline{(\psi_R)^c}\, \Delta_R\, \psi_R \text{ where } \Delta_R \text{ has } I_L = 0 \quad I_R = 1$$

If we now take the fields in the general free Lagrangian (5.6) to be the neutrino fields in these Yukawa couplings, we see that once the Higgs fields acquire vacuum expectation values, there will be Dirac and Majorana neutrino mass terms with:

$$M_D \sim \langle\Phi\rangle$$

$$M_L \sim \langle\Delta_L\rangle$$

$$M_R \sim \langle\Delta_R\rangle$$

Let us now compare the three parameters M_L, M_D and M_R:

The expectation value $<\Delta_L>$ of the $I_L = 1$ Higgs field affects the ρ parameter of the neutral current neutrino scattering, which is measured to be very close to 1. If we take $\rho = 1$ as a phenomenological constraint, it means that $<\Delta_L>$, hence M_L, vanishes.

Turning to M_D, we note that in the left-right symmetric model (as in most reasonable models), the same spontaneous symmetry breaking which leads to the neutrino Dirac mass also leads to the Dirac masses of the charged fermions. Now, charged particles cannot have Majorana masses, since, for example, a mass term such as:

$$\overline{(e_L)^c}\, e_L = e_L^t\, \Omega\, e_L$$

would violate electric charge conservation. Therefore, for the charged fermions, the Dirac masses are the physical masses. Thus, we expect the neutrino Dirac mass M_D to be of the order of the "observed" masses of the related charged lepton and quarks.

Finally, since $<\Delta_R>$ gives W_R its very large mass, we expect that $M_R \sim <\Delta_R>$ is also very large, and, in particular, that

$$M_R >> M_D$$

Hence, the neutrino mass matrix of Eq. (5.8) is of the following type:

$$[M] = \begin{bmatrix} 0 & M_D \\ M_D & M_R \end{bmatrix} \tag{5.11}$$

with M_R dominating M_D.

The eigenvalues are then:

$$M_N \simeq M_R$$

and (5.12)

$$M_{\nu'} \simeq -\frac{M_D^2}{M_R}$$

In terms of the fields f and F of Eqs. (5.7) and (5.9), the corresponding eigenvector fields are:

$$N \simeq F + \frac{M_D}{M_R} f$$

and (5.13)

$$\nu' \simeq f - \frac{M_D}{M_R} F \tag{5.14}$$

In order to get a positive mass for the light neutrino, we take the physical neutrino field to be the field ν related to ν' by:

$$\nu = \gamma_5 \nu' \tag{5.15}$$

Note that $\bar{\nu}' \nu' = -\bar{\nu}\nu$, but $\bar{\nu}' \gamma_\mu \partial_\mu \nu' = +\bar{\nu}\gamma_\mu \partial_\mu \nu$. Thus, in terms of ν and N, the Lagrangian of Eq. (5.10) becomes:

$$-L = [\bar{\nu}\gamma_\mu \partial_\mu \nu + M_\nu \bar{\nu}\nu] + [\bar{N}\gamma_\mu \partial_\mu N + M_N \bar{N} N] \tag{5.16}$$

with $$M_\nu \simeq + \frac{M_D^2}{M_R} \text{ and } M_N \simeq M_R \tag{5.17}$$

If $M_D \sim M_{q \text{ or } \ell}$, a typical quark or charged lepton mass, as we expect, then from Eq. (5.17) we see that

$$\boxed{M_\nu \cdot M_N = M_{q \text{ or } \ell}^2} \tag{5.18}$$

This is the famous *see-saw relation.*[1]

If there is indeed a heavy neutral lepton N, as Eqs. (5.16) and (5.17) suggest, then one can understand from the see-saw relation why the ordinary neutrino is light. Conversely, if one believes in the mechanism that we have described, the fact that the neutrino is light suggests that we should find sooner or later a heavy neutral lepton N.

3 - PROPERTIES AND INTERACTIONS OF THE MASS EIGENSTATES

Let us first actually prove that the neutrinos ν and N are Majorana particles.

We have:
$$N = F + \frac{M_D}{M_R} f$$
$$\nu' = f - \frac{M_D}{M_R} F$$

where
$$f = \frac{\psi_L + (\psi_L)^c}{\sqrt{2}} \quad , \quad F = \frac{\psi_R + (\psi_R)^c}{\sqrt{2}}$$

Now, if one charge-conjugates some field ψ twice, one gets the original spinor back:

$$(\psi^c)^c = \Omega\, \overline{\psi^c}^t = \Omega\, [\psi^t \Omega]^t = \Omega\, \Omega^t \psi = \psi$$
(see Table 3)

We then have:

$$f^c = f \text{ and } F^c = F$$

Therefore:

$$\nu'^c = \nu' \text{ and } N^c = N \tag{5.19}$$

and:

$$\nu^c = (\gamma_5 \nu')^c = \Omega\, \gamma_4 (\gamma_5 \nu')^* = \Omega\, \gamma_4 \gamma_5^* \nu'^*$$
γ_5 is a real matrix
$$= \Omega\, \gamma_4 \gamma_5 \nu'^* = -\gamma_5 \nu'^c = -\gamma_5 \nu'$$

Thus:

$$\nu^c = -\nu \tag{5.20}$$

Now, we saw in Chapter III.2(c) that if a field is proportional to its charge conjugate, then it describes a Majorana (CPT self-conjugate) particle. Thus:

ν and N are Majorana fermions.

Now, let us see how ν and N couple to the W bosons. To do so, we write the charged currents in terms of the eigenfields.

The assignment of ψ_R and e_R to an I_R doublet means that the W_R boson is coupled to:

$$\frac{1}{\sqrt{2}}\bar{e}_R\gamma_\mu\psi_R \tag{5.21}$$

From Eq. (5.7), this equals $\bar{e}_R\gamma_\mu F_R$.

Similarly, the W_L boson is coupled to:

$$\frac{1}{\sqrt{2}}\bar{e}_L\gamma_\mu\psi_L = \bar{e}_L\gamma_\mu f_L$$

Expressing f in terms of the eigenfields ν' and N, one obtains:

$$\bar{e}_L\gamma_\mu f_L \cong \bar{e}_L\gamma_\mu[\nu' + VN]_L$$

where
$$V = \frac{M_D}{M_R}\,.$$

Now
$$\nu'_L = \frac{1+\gamma_5}{2}\nu' = \frac{1+\gamma_5}{2}\gamma_5\nu = \frac{1+\gamma_5}{2}\nu = \nu_L$$

The left-handed charged weak current can then be written:

$$\frac{1}{\sqrt{2}}\bar{e}_L\gamma_\mu\psi_L = \bar{e}_L\gamma_\mu[\nu_L + VN_L] \tag{5.22}$$

We have already talked about neutrino *flavor* mixing, in which the neutrinos ν_e, ν_μ, and ν_τ in the different generations are different linear combinations of a common set of neutrino mass eigenstates. Now, in Eq. (5.22) and its analogue for the right-handed current, we encounter a different kind of neutrino mixing, which we might call *left-right* neutrino mixing. Here, the neutrinos ψ_L and ψ_R of *one* generation which couple, respectively, to W_L and W_R, are different linear combinations of the two mass eigenstates ν and N. The quantity V is a mixing matrix element. Note that

$$V^2 = \frac{M_D^2}{M_R^2} = \frac{M_D^2}{M_R}\,\frac{1}{M_R} = \frac{M_\nu}{M_N} \tag{5.23}$$

That is, in this particular gauge model the mixing matrix element and the neutral lepton masses are simply related.

From the ν and N couplings to the W bosons one can deduce a few experimental consequences.

For example, in doing deep inelastic electron-proton scattering at HERA with a left-handed electron beam, one may produce the heavy lepton N via the adjoint of the current (5.22), but only with a probability $V^2 = M_\nu/M_N$, which is very small. Working with a right-handed or unpolarized electron beam, one might produce N via the adjoint of the current (5.21) and W_R exchange, but the rate for this would also be small due to the large mass of W_R.

There can be a spectacular phenomenological consequence of models of the type we have just discussed in neutrinoless double β decay. These models predict the existence of two Majorana particles, ν and N, whose exchange in the diagrams of Chapter IV.2 should indeed produce this decay. However, suppose that the neutrino mass matrix is of the form (5.11), with $M_R >> M_D$, but that (contrary to our arguments) M_R and M_D are actually small enough that both M_ν and M_N are far below 10 MeV, the typical momentum transfer in double β decay.[34] Then (see Chapter IV.2) the double β decay amplitude coming from the diagram in which the intermediate W bosons are both W_L is proportional to the effective mass:

$$M_{\text{eff}} = \sum_m \lambda_m\, U_{em}^2\, M_m$$

In this sum, our mass eigenstates ν and N have creation phase factors

$\lambda_\nu = -1$ and $\lambda_N = +1$. This follows from Eqs. (5.19) and (5.20) and the fact that any Majorana field χ^M obeys $(\chi^M)^c = \lambda^* \chi^M$. The mixing matrix elements are given by Eqs. (5.22) and (5.23). Thus,

$$M_{\text{eff}} = \underbrace{(-1)(1)^2 M_\nu}_{\nu} + \underbrace{(+1)\frac{M_\nu}{M_N} M_N}_{N} = 0 \tag{5.24}$$

From the fact that they contribute to M_{eff} with opposite signs, ν and N evidently have opposite CP parities (cf. Eq. (4.14)). What is more, we see that they cancel each other *completely* in M_{eff}! This model illustrates how M_{eff} can, in a natural way, be anomalously small compared to the neutrino masses. In consequence, the intermediate $W_L W_L$ amplitude for neutrinoless double β decay is anomalously small.

To be sure, in this model there is also an amplitude for double β decay from the diagram in which the two intermediate W bosons are a W_L and a W_R. However, owing to the W_R propagator, this amplitude is suppressed by a factor:

$$\left[\frac{M_{W_L}}{M_{W_R}}\right]^2 \lesssim \left[\frac{83\text{ GeV}}{1.6\text{ TeV}}\right]^2 \simeq \frac{1}{400}$$

Thus, in some gauge theories the rate for neutrinoless double β decay is very small, even though massive Majorana neutrinos are present.

4 – NEW NEUTRAL LEPTONS IN GAUGE THEORIES

As the left-right symmetric model illustrates, numerous gauge models predict not only nonzero masses for the known neutrinos, but also additional, so far unknown, massive neutral leptons. Let us give some examples.

i) Left-right symmetric models

The one-generation left-right symmetric model we discussed can, of course, be generalized. For each generation, there are two neutrinos, one being light, the other heavy. On top of the mixing between the two neutral leptons in each generation, one can have mixing among the generations.

ii) O(18) Unification

In an attempt to unify the generations, a grand unified theory based on the group O(18) has been developed.[44] In this model, all three known generations, plus many other particles we haven't seen yet, are put into one big family. This family has 256 members (some of which are extremely heavy) and in particular contains four neutrinos

$$\nu_e \quad \nu_\mu \quad \nu_\tau \quad \nu_{\tau'}$$

with left-handed coupling to W (83 GeV), and four associated heavy neutrinos with right-handed coupling to the same W. All eight of these neutrinos have masses less than $M_Z/2$, and so should appear in the $Z°$ decay.

iii) Superstring theories

Some of the superstring theories suggest that the low-energy world (far below the Planck mass $M_P = 10^{19}$ GeV) is described by the grand unified group E_6. The quarks and the leptons of each generation are put in the fundamental representation of this group, a 27-plet. In this model there are 5 Majorana neutrinos per generation.[45,46]

EXERCISES

1 – In the left-right symmetric model we have discussed, the W_L boson couples to the left-handed current of Eq. (5.22). More accurately normalized, this current may be written as:

$$\bar{e}_L \gamma_\mu [\cos\theta\, \nu_L + \sin\theta\, N_L]$$

Here θ is a mixing angle which describes the linear combination of mass eigenstates ν and N which couples to W_L. Show that so long as the real, symmetric mass matrix $[M]$ of Eq. (5.8) has the form

$$[M] = \begin{bmatrix} 0 & X \\ X & A \end{bmatrix}$$

with a zero in the upper left corner, θ will be related to the masses of ν and N by:

$$\tan^2 \theta = \frac{M_\nu}{M_N}$$

2 – The fact that the mass eigenstates ν and N contribute to M_{eff} of Eq. (5.24) with opposite sign implies that they have opposite CP parities:

$$\tilde{\eta}_{\text{CP}}(\nu) = -\tilde{\eta}_{\text{CP}}(N)$$

Prove directly that this is indeed true using the relation $\tilde{\eta}_{\text{CP}} = \eta_{\text{CP}}\,\lambda$ for any Majorana particle, the relation (4.10) between the phases of the mixing matrix elements and the phases $\eta_{\text{CP}}(\nu_m)$, and the fact that $\lambda_\nu = -1$ while $\lambda_N = +1$.

CHAPTER

SIX

CONCLUSION

Massive neutrinos are fascinating physical objects. They are expected to undergo neutrino oscillations in vacuum, and perhaps matter-enhanced flavor transitions in the sun. Both of these phenomena are very interesting examples of the workings of quantum mechanics, and the second may be the solution to the solar neutrino puzzle.

If, besides being massive, neutrinos are also their own antiparticles, then they possess a number of unusual properties, such as imaginary CP parities, and vanishing magnetic and electric dipole moments.

In addition to being very interesting objects, massive neutrinos are also more natural than massless ones from the standpoint of the grand unified theories. This fact is one more good reason to carry out experimental searches for evidence of neutrino mass. We look forward to learning what the present and future searches will reveal.

SOLUTIONS TO THE EXERCISES

1 - CHAPTER II

1 - The neutrino is born as $\nu_\ell = \sum_m U_{\ell m}\, \nu_m$, so that:

$$\psi(x, t = 0) = \sum_m U_{\ell m}\, \nu_m\, e^{ip_m x}$$

After a time t, this evolves into:

$$\psi(x, t) = \sum_m U_{\ell m} \nu_m\, e^{ip_m x}\, e^{-iE_\nu t} \cong \sum_m U_{\ell m} \nu_m e^{i(E_\nu - M_m^2/2E_\nu)x}\, e^{-iE\nu t}$$

if we use $p_m \cong E_\nu - \dfrac{M_m^2}{2E_\nu}$. Then:

$$\psi(x, x) \cong \sum_m U_{\ell m} \nu_m\, e^{-i(M_m^2/2E_\nu)x} = \sum_{\ell'} \left[\sum_m U_{\ell m} e^{-i(M_m^2/2E_\nu)x}\, U^*_{\ell' m} \right] \nu_{\ell'}$$

2 - The U_{em} are constrained by the condition:

$$\sum_{m=1}^{N} U_{em}^2 = 1$$

Let us use the method of Lagrange multipliers, and define:

$$F = P_{ee} + \lambda\left[\sum_m U_{\mathrm{em}}^2 - 1\right]$$

Here λ is the Lagrange multiplier. Requiring that $\partial F/\partial U_{em} = 0$ at the minimum, we find that at that point:

$$U_{em}^2 = -\frac{\lambda}{2}, \quad m = 1, \ldots, N$$

That is, all U_{em}^2 are equal at the minimum, and since $\sum_m U_{em}^2 = 1$, they are all equal to $\frac{1}{N}$. Then:

$$[P_{ee}]_{\min} = \left[\sum_{m=1}^{N} U_{em}^4\right]_{\min} = N\left[\frac{1}{N}\right]^2 = \frac{1}{N}$$

For the fluxes of two flavors of neutrino, ℓ_1 and ℓ_2, to be reduced by $\frac{1}{N}$, we must have:

$$P_{\ell\ell} = \sum_{m=1}^{N} U_{\ell m}^4 = \frac{1}{N}$$

for both $\ell = \ell_1$ and $\ell = \ell_2$. From the previous considerations, this means that in both the ℓ_1 and ℓ_2 columns of the U matrix, every entry must be $\pm\frac{1}{\sqrt{N}}$. But then, if N is odd, it is impossible to satisfy the orthogonality condition:[47]

$$\sum_{m=1}^{N} U_{\ell_1 m} U_{\ell_2 m} = 0$$

3 – From Eq. (2.13), the region where $\sin^2 2\theta_M > \frac{1}{2}$ has a width:

$$\Delta\left(\frac{L_v}{L_o}\right) = 2\sin 2\theta_v$$

Since L_v/L_o is proportional to $N_e(r)$:

$$\Delta\left(\frac{L_v}{L_o}\right) = \frac{L_v}{L_o}\frac{\Delta N_e}{N_e} \simeq \frac{L_v}{L_o}\frac{1}{N_e}\left[-\frac{dN_e}{dr}\right]\Delta r$$

(dN_e/dr is negative).

Thus:

$$2\sin 2\theta_v \simeq \frac{L_v}{L_o}\frac{1}{N_e}\left[-\frac{dN_e}{dr}\right]\Delta r$$

The oscillation length in matter near the critical point where $L_v/L_o = \cos 2\theta_v$ is:

$$L_M = \frac{L_v}{\sin 2\theta_v} = \frac{L_o}{\tan 2\theta_v}$$

Thus, requiring that $\Delta r > L_M$, we have that:

$$2\sin 2\theta_v \gtrsim \cos 2\theta_v \frac{1}{N_e}\left[-\frac{dN_e}{dr}\right]\frac{L_o}{\tan 2\theta_v}$$

Using $L_o = 2\pi/(\sqrt{2}\,G_F N_e)$, and disregarding numerical constants of order π, this becomes:[48]

$$\tan^2 2\theta_v \gtrsim \left[\frac{1}{G_F N_e^2}\left(-\frac{dN_e}{dr}\right)\right]_{r=r_{\text{crit}}}$$

2 – CHAPTER III

1 – If C is conserved in the process

$$\gamma + \gamma \rightarrow \tilde{\gamma} + \tilde{\gamma}$$

then:

$$C(\tilde{\gamma}\tilde{\gamma}) = \tilde{\eta}_C^2(\tilde{\gamma})$$
$$= C(\gamma\gamma) = (-1)^2 = +1$$

Hence

$$\tilde{\eta}_C(\tilde{\gamma}) = \pm 1$$

2 – The J of the parent toponium state is 2, so the final $\tilde{\gamma}\tilde{\gamma}$ states allowed by Fermi statistics are:

$$^1D_2,\ ^3F_2,\ ^3P_2$$

Now, P conservation in the decay implies that if L is the orbital angular momentum of the $\tilde{\gamma}\tilde{\gamma}$ final state:

$$P\,(\tilde{\gamma}\tilde{\gamma}\text{ final state}) = \tilde{\eta}_P^2(\tilde{\gamma})\,(-1)^L = (\pm\, i)^2(-1)^L$$
$$= P(^3P_2\text{ toponium state}) = +1$$

Hence, L must be *odd*.
Thus, the P-allowed final states are $^3F_2$ and $^3P_2$, but if P is violated, the $^1D_2$ state is allowed as well.

3 – Let $|\,\nu_i\rangle \equiv |\,\nu^M(\mathbf{p}_i, s_i)\rangle$, and similarly for $|\,\nu_f\rangle$. For $k = 1, 2$, or 3, J_k^{EM} is a hermitian operator. Since ζ is antiunitary:

$$\langle \nu_f | J_k^{EM} | \nu_i \rangle = \langle J_k^{EM} \nu_f | \nu_i \rangle = \langle \zeta \nu_i | \zeta J_k^{EM} \nu_f \rangle$$
$$= \langle \zeta \nu_i | \zeta J_k^{EM} \zeta^{-1} | \zeta \nu_f \rangle$$

Using Eq. (3.9) and the fact that J_k^{EM} is CPT-odd, this is:

$$= \tilde{\eta}_\zeta^{s_i *} \tilde{\eta}_\zeta^{s_f} \langle \nu^M(\mathbf{p}_i, -s_i) | -J_k^{\mathrm{EM}} | \nu^M(\mathbf{p}_f, -s_f) \rangle$$

Using Eq. (3.10), $\tilde{\eta}_\zeta^{s_i *} \tilde{\eta}_\zeta^{s_f} = (-1)^{s_i - s_f}$, so this is:

$$= -(-1)^{s_i - s_f} \langle \nu^M(\mathbf{p}_i, -s_i) | J_k^{\mathrm{EM}} | \nu^M(\mathbf{p}_f, -s_f) \rangle$$

Turning to J_4^{EM}, in the metric we are using $J_4^{\mathrm{EM}} = i\rho$, where the charge-density operator ρ is Hermitian and CPT-odd (since φ is CPT-odd, and $H^{EM} = \mathbf{J}^{\mathrm{EM}} \cdot \mathbf{A} - \rho\, \varphi$ must be CPT-even). The analysis we have just gone through for J_k^{EM} then applies to ρ as well, so that:

$$\langle \nu_f | \rho | \nu_i \rangle = -(-1)^{s_i - s_f} \langle \nu^M(\mathbf{p}_i, -s_i) | \rho | \nu^M(\mathbf{p}_f, -s_f) \rangle$$

Multiplying this relation by i and combining it with the previous one for J_k^{EM}, we have finally that for $\mu = 1, 2, 3, 4$:

$$\langle \nu^M(\mathbf{p}_f, s_f) | J_\mu^{\mathrm{EM}} | \nu^M(\mathbf{p}_i, s_i) \rangle = -(-1)^{s_i - s_f} \langle \nu^M(\mathbf{p}_i, -s_i) | J_\mu^{\mathrm{EM}} | \nu^M(\mathbf{p}_f, -s_f) \rangle$$

4 – Applying the expression for $\langle | J_\mu^{\mathrm{EM}} | \rangle$ stated in the problem to both sides of the CPT constraint of problem III.3, we have:

$$\langle \nu^M(\mathbf{p}_f, s_f) | J_\mu^{\mathrm{EM}} | \nu^M(\mathbf{p}_i, s_i) \rangle = i\bar{u}(\mathbf{p}_f, s_f) \left[F\gamma_\mu + G(q^2 \gamma_\mu - \not{q} q_\mu)\gamma_5 \right.$$
$$\left. + M \sigma_{\mu\nu} q_\nu + E i \sigma_{\mu\nu} q_\nu \gamma_5 \right] u(\mathbf{p}_i, s_i) \tag{i}$$

$$= -(-1)^{s_i - s_f} i\bar{u}(\mathbf{p}_i, -s_i) [F\gamma_\mu + G(q^2 \gamma_\mu - \not{q} q_\mu)\gamma_5 - M \sigma_{\mu\nu} q_\nu - E i \sigma_{\mu\nu} q_\nu \gamma_5]$$
$$\times\, u(\mathbf{p}_f, -s_f)$$

Here we have used the fact that on the right hand side (RHS) of the CPT constraint, the momentum transfer $p_{\text{initial}} - p_{\text{final}} = p_f - p_i = -q$.

From Table 2:

$$u(\mathbf{p}, -s) = (-1)^{s-1/2}\gamma_5 v(\mathbf{p}, s) = (-1)^{s-1/2}\gamma_5\gamma_2 u^*(\mathbf{p}, s)$$
$$= (-1)^{s-1/2}\gamma_5 \Omega \bar{u}^t(\mathbf{p}, s)$$

Hence

$$\bar{u}(\mathbf{p}, -s) = (-1)^{s-1/2} u^t(\mathbf{p}, s)\Omega^{-1}\gamma_5$$

(see Table 3 for properties of Ω)

Thus, the RHS of Eq. (i) above is:

$$\text{RHS} = -(-1)^{s_i - s_f}(-1)^{s_i-1/2}(-1)^{s_f-1/2} i u^t(\mathbf{p}_i, s_i)\Omega^{-1}\gamma_5$$
$$\times [F\gamma_\mu + G(q^2\gamma_\mu - \not{q}q_\mu)\gamma_5 - M\sigma_{\mu\nu}q_\nu - Ei\sigma_{\mu\nu}q_\nu\gamma_5]\gamma_5\Omega\bar{u}^t(\mathbf{p}_f, s_f)$$
$$= -iu^t(\mathbf{p}_i, s_i)\Omega^{-1}[-F\gamma_\mu - G(q^2\gamma_\mu - \not{q}q_\mu)\gamma_5$$
$$- M\sigma_{\mu\nu}q_\nu - Ei\sigma_{\mu\nu}q_\nu\gamma_5]\Omega\bar{u}^t(\mathbf{p}_f, s_f)$$

Using the properties of the various gamma matrices under charge conjugation, this is:

$$= -i\,u^t(\mathbf{p}_i, s_i)\{F\gamma_\mu^t - G[(q^2\gamma_\mu - \not{q}q_\mu)\gamma_5]^t$$
$$+ M[\sigma_{\mu\nu}q_\nu]^t + E[i\,\sigma_{\mu\nu}q_\nu\gamma_5]^t\}\bar{u}^t(\mathbf{p}_f, s_f)$$
$$= -i\,\bar{u}(\mathbf{p}_f, s_f)[F\gamma_\mu - G(q^2\gamma_\mu - \not{q}q_\mu)\gamma_5$$
$$+ M\sigma_{\mu\nu}q_\nu + Ei\sigma_{\mu\nu}q_\nu\gamma_5]\,u(\mathbf{p}_i, s_i)$$

Comparing this expression to that in the middle of Eq. (i), we see that the form factors F, M, and E must vanish.

5 - Let CP $\equiv Z$.

Dirac case

$$\Psi = \Psi^D = \sum_{\mathbf{p},s} \sqrt{\frac{M}{E(\mathbf{p})V}}[f(\mathbf{p},s)u(\mathbf{p},s)\,e^{ipx} + \bar{f}^\dagger(\mathbf{p},s)v(\mathbf{p},s)e^{-ipx}]$$

$$Z\Psi^D(\mathbf{x},t)Z^{-1} = \sum_{\mathbf{p},s} \sqrt{\frac{M}{E(\mathbf{p})V}}\,[Zf(\mathbf{p},s)Z^{-1}\,u(\mathbf{p},s)e^{ipx} + Z\bar{f}^\dagger(\mathbf{p},s)\,Z^{-1}v(\mathbf{p},s)e^{-ipx}]$$

$$\eta_Z^*\,\gamma_4\gamma_2\,[\Psi^D(-\mathbf{x},t)]^* = \eta_Z^*\,\gamma_4\sum_{\mathbf{p},s} \sqrt{\frac{M}{E(\mathbf{p})V}}\,[\bar{f}(\mathbf{p},s)u(\mathbf{p},s)e^{+i(\mathbf{p}\cdot(-\mathbf{x})-E(\mathbf{p})t)} + f^\dagger(\mathbf{p},s)v(\mathbf{p},s)e^{-i(\mathbf{p}\cdot(-\mathbf{x})-E(\mathbf{p})t)}]$$

$$= \eta_Z^*\sum_{\mathbf{p},s} \sqrt{\frac{M}{E(\mathbf{p})V}}\,[\bar{f}(\mathbf{p},s)u(-\mathbf{p},s)e^{i(\mathbf{p}\cdot(-\mathbf{x})-E(\mathbf{p})t)} - f^\dagger(\mathbf{p},s)v(-\mathbf{p},s)e^{-i(\mathbf{p}\cdot(-\mathbf{x})-E(\mathbf{p})t)}]$$

Here we have used first the relations

$$\gamma_2\,u^*(\mathbf{p},s) = v(\mathbf{p},s), \quad \gamma_2\,v^*(\mathbf{p},s) = u(\mathbf{p},s)$$

from Table 2, and then the relations

$$\gamma_4\,u(\mathbf{p},s) = u(-\mathbf{p},s), \quad \gamma_4\,v(\mathbf{p},s) = -v(-\mathbf{p},s)$$

Changing the summation label from $\mathbf{p}$ to $-\mathbf{p}$, we have:

$$\eta_Z^*\,\gamma_4\,\gamma_2\,[\Psi^D(-\mathbf{x},t)]^* = \eta_Z^*\sum_{\mathbf{p},s} \sqrt{\frac{M}{E(\mathbf{p})V}}[\bar{f}(-\mathbf{p},s)\,u(\mathbf{p},s)e^{ipx} - f^\dagger(-\mathbf{p},s)\,v(\mathbf{p},s)e^{-ipx}]$$

Equating this expression to that for $Z\Psi^D(\mathbf{x}, t)Z^{-1}$, we conclude that:

$$Zf(\mathbf{p}, s)Z^{-1} = \eta_Z^* \bar{f}(-\mathbf{p}, s), \quad Z\bar{f}^\dagger(\mathbf{p}, s)Z^{-1} = -\eta_Z^* f^\dagger(-\mathbf{p}, s)$$

Hermitian conjugating the first of these relations (using $Z^{-1} = Z^\dagger$), we find that:

$$Zf^\dagger(\mathbf{p}, s)Z^{-1} = \eta_Z \bar{f}^\dagger(-\mathbf{p}, s)$$

Applying this relation to the vacuum, which we define, of course, to be invariant under CP, we obtain:

$$Z|f(\mathbf{p}, s)\rangle = \eta_Z |\bar{f}(-\mathbf{p}, s)\rangle$$

Similarly, applying the CP transformation law derived above for $\bar{f}^\dagger$ to the vacuum, we obtain:

$$Z|\bar{f}(\mathbf{p}, s)\rangle = -\eta_Z^* |f(-\mathbf{p}, s)\rangle$$

Notice that:

$$Z^2|f(\mathbf{p}, s)\rangle = -|f(\mathbf{p}, s)\rangle$$

Majorana case

The results for this case can be obtained from those for the Dirac case by substituting $\bar{f}^\dagger = \lambda f^\dagger$, but it is instructive to be more explicit. We have for this case:

$$\Psi = \Psi^M = \sum_{\mathbf{p}, s} \sqrt{\frac{M}{E(\mathbf{p})V}} \left[f(\mathbf{p}, s)u(\mathbf{p}, s)e^{ipx} + \lambda f^\dagger(\mathbf{p}, s)v(\mathbf{p}, s)e^{-ipx} \right]$$

For $Z\Psi^M(\mathbf{x}, t)\, Z^{-1}$ we have an expression analogous to that for the Dirac case. Furthermore, since $\gamma_2 \Psi^{M*} = \lambda^* \Psi^M$ (see lines preceding Eq. (3.29)):

$$\eta_Z^* \gamma_4 \gamma_2 [\Psi^M(-\mathbf{x}, t)]^* = \eta_Z^* \lambda^* \sum_{\mathbf{p}, s} \sqrt{\frac{M}{E(\mathbf{p})V}}$$

$$[f(-\mathbf{p}, s) u(\mathbf{p}, s) e^{ipx} - \lambda f^\dagger(-\mathbf{p}, s) v(\mathbf{p}, s) e^{-ipx}]$$

Here we have again used the relations for $\gamma_4 u$ and $\gamma_4 v$ quoted earlier, and changed the summation label from $\mathbf{p}$ to $-\mathbf{p}$.

Equating the last expression to that for $Z\Psi^M(\mathbf{x}, t)Z^{-1}$, we find that:

$$Zf(\mathbf{p}, s)Z^{-1} = (\eta_Z \lambda)^* f(-\mathbf{p}, s) \; and \; Zf^\dagger(\mathbf{p}, s)Z^{-1} = -(\eta_Z \lambda)^* f^\dagger(-\mathbf{p}, s)$$

Hermitian conjugating the first of these relations yields:

$$Zf^\dagger(\mathbf{p}, s)Z^{-1} = (\eta_Z \lambda) f^\dagger(-\mathbf{p}, s)$$

Note that we now have two expressions for the effect of CP on $f^\dagger(\mathbf{p}, s)$. Defining $\tilde{\eta}_Z \equiv \eta_Z \lambda$, we see that the consistency of these two expressions requires that:

$$\tilde{\eta}_Z = -\tilde{\eta}_Z^*$$

Applying either of the expressions for $Zf^\dagger Z^{-1}$ to the vacuum yields:

$$Z|f(\mathbf{p}, s)\rangle = \tilde{\eta}_Z |f(-\mathbf{p}, s)\rangle$$
$$\text{with } \tilde{\eta}_Z = \eta_Z \lambda = -\tilde{\eta}_Z^*$$

3 - CHAPTER IV

1 - The couplings g and propagators of the W bosons in the diagram at the beginning of Chapter IV.2 contribute a factor $[g^2/(83 \text{ GeV})^2]^2$ to this diagram. Thus, from the expression immediately preceding Eq. (4.7) and the definition of M_{eff}, the amplitude A_ν for this diagram, summed over all ν_m exchanges, has the form:

$$A_\nu \sim (\text{Nucl}) \left(\frac{g}{83 \text{ GeV}}\right)^4 \frac{M_{\text{eff}}}{(10 \text{ MeV})^2} A_{uv} \equiv F M_{\text{eff}}$$

Here "Nucl" is a nuclear matrix element, A_{uv} is a factor involving the u and v spinors for the electrons and gamma matrices, and we have replaced the $1/q^2$ in the expression before Eq. (4.7) by $1/(10\ \text{MeV})^2$, since 10 MeV is a typical value of q in $\beta\beta_{ov}$.

The experimental upper bound on M_{eff} is obviously obtained by dividing the experimental upper bound on the $\beta\beta_{ov}$ amplitude by a theoretical estimate for the factor F.

Now, suppose that the $\beta\beta_{ov}$ amplitude is actually dominated by A_N, not A_ν. Since A_N involves currents of the *same* handedness at both lepton vertices, it will be proportional to M_N, just as if all the N couplings were left-handed. However, since $M_N >> 10$ MeV, the denominator of the N-exchange propagator will be dominated by M_N^2, not $q^2 \cong (10\ \text{MeV})^2$. Thus, modifying A_ν appropriately, we find that:

$$A_N \sim (\text{Nucl}) \left(\frac{g}{M_{W_R}}\right)^4 \frac{M_N}{M_N^2} A'_{uv}$$

where A'_{uv} is very similar to A_{uv}.

The existing bound $M_{\text{eff}} \lesssim 1$ eV means that:

$$\frac{A\,[\beta\beta_{ov}]\,\big|_{\text{Experimental}}}{F} \lesssim 1\ \text{eV}$$

If A_N dominates $\beta\beta_{ov}$, this bound implies that:

$$\left(\frac{83\ \text{GeV}}{M_{W_R}}\right)^4 \frac{(10\ \text{MeV})^2}{M_N} \lesssim 1\ \text{eV}$$

If $M_{W_R} = 1.6$ TeV, then:

$$M_N \gtrsim 0.7\ \text{GeV}$$

2 – Suppose first that ν is a Majorana particle. Then $\bar{\nu}\,\gamma_\mu \nu = 0$. When H_{NC} causes the decay $Z°(\varepsilon_\mu) \rightarrow \nu(\mathbf{p}_1, s_1)\,\nu(\mathbf{p}_2, s_2)$, either the field $\bar{\nu}$ creates

$\nu(\mathbf{p}_1, s_1)$ and ν creates $\nu(\mathbf{p}_2, s_2)$, or the other way around. From the plane-wave expansions for ν and $\bar{\nu}$, the amplitude for the decay is, apart from an overall phase:

$$A_M = \varepsilon_\mu(\bar{u}_1\gamma_\mu\gamma_5 v_2 - \bar{u}_2\gamma_\mu\gamma_5 v_1)$$

Here $u_1 \equiv (\mathbf{p}_1, s_1)$, $v_1 = v(\mathbf{p}_1, s_1)$, and similarly for u_2, v_2.
Now, from Table 2:

$$v(\mathbf{p}, s) = \gamma_2\, u^*(\mathbf{p}, s) = \Omega\, \bar{u}^t(\mathbf{p}, s)$$

From the second line of Table 3, this implies:

$$\bar{u}(\mathbf{p}, s) = -v^t(\mathbf{p}, s)\,\Omega^{-1}$$

Now since, as we have shown,

$$\Omega^{-1}\,\gamma_\mu\gamma_5\,\Omega = (\gamma_\mu\gamma_5)^t$$

these relations imply that:

$$-\bar{u}_2\,\gamma_\mu\gamma_5\,v_1 = +\bar{u}_1\,\gamma_\mu\gamma_5\,v_2$$

Hence:

$$A_M = 2\,\bar{u}_1\,\not{\varepsilon}\,\gamma_5\,v_2 \qquad \text{(i)}$$

Next, suppose ν is a Dirac particle. Then, using standard procedures, we find from H_{NC} that the amplitude for the decay $Z°(\varepsilon_\mu) \rightarrow \nu(\mathbf{p}_1, s_1)$ $\bar{\nu}(\mathbf{p}_2, s_2)$ is:

$$A_D = \bar{u}_1\,\not{\varepsilon}\,(1 + \gamma_5)\,v_2 \qquad \text{(ii)}$$

Relativistic limit

From the discussion in Chapter III.2.d, it follows that when the outgoing neutrinos are highly relativistic, A_D vanishes due to the $(1 + \gamma_5)$

unless the ν is left-handed (L) and the $\bar{\nu}$ is right-handed (R). For this one allowed helicity configuration, $(1 + \gamma_5)$ may be replaced by 2. Thus, if $\mathbf{p}$ is the momentum of the ν in the $Z°$ rest frame, and we neglect factors that are common to the Dirac and Majorana cases, the angle-integrated $Z°$ decay rate in the Dirac case is:

$$\Gamma_D = \int |2\,\bar{u}(\mathbf{p}, L)\, \not{\epsilon}\, v(-\mathbf{p}, R)|^2\, d\Omega_{\mathbf{p}}$$

Turning to the Majorana case, we note from the form of A_M that when the outgoing neutrinos are highly relativistic, they must be of opposite helicity. For example, if the neutrino described by v_2 is right-handed, then:

$$A_M = 2\,\bar{u}_1\, \not{\epsilon}\, \gamma_5\, v_2 = 2\,\bar{u}_1\, \not{\epsilon}\, \gamma_5 \frac{1+\gamma_5}{2} v_2 = 2\, \overline{\left[\frac{1+\gamma_5}{2} u_1\right]}\, \not{\epsilon}\, \gamma_5\, v_2$$

Thus, in this example the neutrino described by u_1 must be left-handed or A_M will vanish. Furthermore, so long as the neutrino described by v_2 has some definite helicity, positive or negative, $\gamma_5\, v_2 = \pm\, v_2$. Hence, if $\mathbf{p} = \mathbf{p}_1 = -\mathbf{p}_2$ is the outgoing momentum of one of the neutrinos in the $Z°$ rest frame, the angle-integrated $Z°$ decay rate in the Majorana case is:

$$\Gamma_M = \frac{1}{2}\int |2\,\bar{u}(\mathbf{p}, L)\, \not{\epsilon}\, v(-\mathbf{p}, R)|^2 + |2\,\bar{u}(\mathbf{p}, R)\, \not{\epsilon}\, v(-\mathbf{p}, L)|^2\, d\Omega_{\mathbf{p}}$$

Here the factor 1/2 must be included because the two outgoing particles are identical.

We see that the first term in the expression for Γ_M is just one-half Γ_D. For the second term, we note that:

$$\begin{aligned} 2\,\bar{u}(\mathbf{p}, R)\, \not{\epsilon}\, v(-\mathbf{p}, L) &= -2\, v^t(\mathbf{p}, R)\, \Omega^{-1}\, \not{\epsilon}\, \Omega\, \bar{u}^t(-\mathbf{p}, L) \\ &= -2\, v^t(\mathbf{p}, R)\, \not{\epsilon}^t\, \bar{u}^t(-\mathbf{p}, L) = 2\,\bar{u}(-\mathbf{p}, L)\, \not{\epsilon}\, v(\mathbf{p}, R) \end{aligned}$$

That is, the second term is the same as the first, except for $\mathbf{p} \to -\mathbf{p}$. Since we are integrating over all directions of **p**, the two terms make equal contributions to Γ_M.

Hence:

$$\underline{\Gamma_M = \Gamma_D}$$

Nonrelativistic limit

When the outgoing neutrinos are nonrelativistic, we expect $^3S_1$ to be the dominant final state in the Dirac case. Since $^3P_1$ is the sole final state in the Majorana case, we then expect that:

$$\frac{\Gamma_M}{\Gamma_D} \sim \left(\frac{p/M}{1}\right)^2 = \beta^2$$

Here β is the speed of the outgoing neutrinos in the $Z°$ rest frame.

Using Eqs. (3.11), (3.13), and (3.15), it is straightforward to show that in the nonrelativistic limit, the amplitudes of Eqs. (i) and (ii) become:

$$A_M = -2\, \boldsymbol{\varepsilon}\cdot(\chi_1^\dagger \boldsymbol{\sigma} \chi_2^c) \times \frac{\mathbf{p}}{M}$$

and

$$A_D = -i\, \boldsymbol{\varepsilon}\cdot\chi_1^\dagger \boldsymbol{\sigma} \chi_2^c$$

Here χ_1 is the Pauli spinor that appears in u_1, and χ_2^c the one that appears in v_2.

To calculate Γ_M and Γ_D, we suppose that the $Z°$ is polarized with $J_Z = 0$, so that $\boldsymbol{\varepsilon} = (0, 0, 1, i0)$. Then:

$$\boldsymbol{\varepsilon}\cdot\boldsymbol{\sigma} \times \mathbf{p} = \sigma_x p_y - \sigma_y p_x = p \sin\theta \begin{bmatrix} 0 & ie^{-i\varphi} \\ -ie^{i\varphi} & 0 \end{bmatrix}$$

and:

$$\boldsymbol{\varepsilon}\cdot\boldsymbol{\sigma} = \sigma_z$$

Here θ, φ are the usual angles specifying the direction of **p**.

We have for the Majorana case:

$$\sum_{\substack{\text{final}\\\text{spins}}} |A_M|^2 = \sum_{\substack{\text{final}\\\text{spins}}} 4\frac{p^2}{M^2}\sin^2\theta \left|\chi_1^\dagger \begin{bmatrix} 0 & ie^{-i\varphi} \\ -ie^{i\varphi} & 0 \end{bmatrix} \chi_2^c\right|^2$$

$$= 8\beta^2 \sin^2\theta$$

(Only like spins contribute.)

For the Dirac case:

$$\sum_{\substack{\text{final}\\\text{spins}}} |A_D|^2 = \sum_{\substack{\text{final}\\\text{spins}}} |\chi_1^\dagger \sigma_z \chi_2^c|^2 = 2$$

(Only opposite spins contribute.)

Integrating over all directions of **p**, and remembering the factor of 1/2 for identical final particles in the Majorana case, we find that:

$$\frac{\Gamma_M}{\Gamma_D} = \frac{4}{3}\beta^2$$

We note that if $\beta << 1$, Γ_M and Γ_D are very different.[49]

4 – CHAPTER V

1 – Since $[M]$ can be diagonalized by a rotation, its eigenfields ν' and N can be written in terms of the basis fields f and F as:

$$\begin{aligned} \nu' &= \cos\theta\, f - \sin\theta\, F \\ N &= \sin\theta\, f + \cos\theta\, F \end{aligned} \qquad \text{(i)}$$

Recall that for one of these fields (ν', say), the physical field ν is γ_5 times the eigenfield, and the mass M_ν of the physical particle is the negative of the eigenvalue. Now, from Eqs. (i), we have:

$$f = \cos\theta\, \nu' + \sin\theta\, N$$

From the discussion before Eq. (5.22), the W_L couples to the current:

$$\bar{e}_L \gamma_\mu f_L$$

Thus, it couples to:

$$\bar{e}_L \gamma_\mu [\cos\theta\, \nu'_L + \sin\theta\, N_L] = \bar{e}_L \gamma_\mu [\cos\theta\, \nu_L + \sin\theta\, N_L]$$

From Eqs. (i), the application of $[M]$ to ν' and N yields (with $M_\nu = -M_{\nu'}$):

$$\begin{bmatrix} 0 & X \\ X & A \end{bmatrix} \begin{bmatrix} \cos\theta \\ -\sin\theta \end{bmatrix} = -M_\nu \begin{bmatrix} \cos\theta \\ -\sin\theta \end{bmatrix}$$

and

$$\begin{bmatrix} 0 & X \\ X & A \end{bmatrix} \begin{bmatrix} \sin\theta \\ \cos\theta \end{bmatrix} = M_N \begin{bmatrix} \sin\theta \\ \cos\theta \end{bmatrix}$$

The top lines of these two relations are:

$$-X\sin\theta = -M_\nu \cos\theta$$

$$X\cos\theta = M_N \sin\theta$$

Dividing the first of these equations by the second, we obtain:

$$\tan^2\theta = \frac{M_\nu}{M_N}$$

2 – The mixing matrix elements $U_{em}(m = \nu, N)$ which appear in the left-handed charged weak current (5.22) are:

$$U_{e\nu} = 1,\ U_{eN} = V = \frac{M_D}{M_R}$$

These U_{em} are both real. Therefore, from Eq. (4.10), the phases $\eta_{CP}(\nu)$ and $\eta_{CP}(N)$ must be equal. Thus:

$$\frac{\tilde{\eta}_{CP}(\nu)}{\tilde{\eta}_{CP}(N)} = \frac{\eta_{CP}(\nu)\,\lambda_\nu}{\eta_{CP}(N)\,\lambda_N} = \frac{\lambda_\nu}{\lambda_N} = -1$$

2 – The mixing matrix elements $U_{em}(m = \nu, N)$ which appear in the left-handed charged weak current (5.22) are:

$$U_{e\nu} = 1, \; U_{eN} = V = \frac{M_D}{M_R}$$

These U_{em} are both real. Therefore, from Eq. (4.10), the phases $\eta_{\text{CP}}(\nu)$ and $\eta_{\text{CP}}(N)$ must be equal. Thus:

$$\frac{\tilde{\eta}_{\text{CP}}(\nu)}{\tilde{\eta}_{\text{CP}}(N)} = \frac{\eta_{\text{CP}}(\nu)\,\lambda_\nu}{\eta_{\text{CP}}(N)\,\lambda_N} = \frac{\lambda_\nu}{\lambda_N} = -1$$

REFERENCES

1. M. Gell-Mann, P. Ramond, and R. Slansky, in *Supergravity*, edited by D. Freedman and P. van Nieuwenhuizen (North Holland, Amsterdam, 1979), p. 315; T. Yanagida, in *Proceedings of the Workshop on Unified Theory and Baryon Number in the Universe*, edited by O. Sawada and A. Sugamoto (KEK, Tsukuba, Japan, 1979); R. Mohapatra and G. Senjanovic, *Phys. Rev. Lett.* **44**, (1980), 912 and *Phys. Rev.* **D23**, (1981) 165.
2. R. Shrock, *Phys. Rev.* **D24**, (1981) 1232.
3. See, for example, R. Prieels, in *Proceedings of the Fourth Moriond Workshop on Massive Neutrinos in Astrophysics and in Particle Physics*, edited by J. Trân Thanh Vân (Editions Frontieres, Gif-sur-Yvette, France, 1984), p. 189, and references therein.
4. R. Abela *et al.*, *Phys. Lett.* **146B**, (1984) 431.
5. S. Boris *et al.*, *Phys. Rev. Lett.* **58**, (1987) 2019.
6. From one of these tritium experiments, an upper limit of 18 eV on the mass of the dominant mass eigenstate in ν_e is reported in M. Fritschi *et al.*, *Phys. Lett.* **173B**, (1986) 485. From another, a limit of 27 eV is reported in J. Wilkerson *et al.*, *Phys. Rev. Lett.* **58**, (1987) 2023. For a review of the experiments bearing on neutrino mass, and further discussion of the associated physics, see F. Boehm and P. Vogel, *Physics of Massive Neutrinos* (Cambridge University Press, Cambridge, 1987).
7. Analyses of the spread in arrival times and the energies of the neutrinos from supernova 1987A have also yielded an upper limit on the mass of the dominant mass eigenstate in ν_e. Interestingly enough, this limit is (20–30) eV, or perhaps smaller, but there appears to be no conclusive, model-independent limit stringent enough to rule out the nonzero result of the Moscow experiment. See, for example, J. Bahcall and S. Glashow, Nature **326**, (1987)

476; D. Arnett and J. Rosner, *Phys. Rev. Lett.* **58**, (1987) 1906; E. Kolb, A. Stebbins, and M. Turner, *Phys. Rev.* **D35**, (1987) 3598.

8. H. Albrecht *et al.* (The ARGUS Collaboration), *Phys. Lett.* **202B**, (1988) 149. A similar limit from a related experiment is reported in S. Abachi *et al.*, *Phys. Rev.* **D35**, (1987) 2880.
9. N. Deshpande and G. Eilam, *Phys. Rev. Lett.* **53**, (1984) 2289.
10. J. Nieves and P. Pal, *Phys. Rev.* **D32**, (1985) 1849.
11. B. Kayser, *Phys. Rev.* **D26**, (1982) 1662, and in *Field Theory in Elementary Particles*, edited by A. Perlmutter (Plenum, New York, 1983), p. 49.
12. B. Kayser, *Phys. Rev.* **D24**, (1981) 110.
13. J. LoSecco *et al.*, *Phys. Rev. Lett.* **54**, (1985) 2299.
14. J. Rich, D. Lloyd Owen, and M. Spiro, *Phys. Rep.* **151**, (1987) 239.
15. S. P. Mikheyev and A. Yu. Smirnov. *Nuovo Cimento* **C9**, (1986) 17.
16. L. Wolfenstein, *Phys. Rev.* **D17**, (1978) 2369.
17. S. P. Rosen and J. Gelb, *Phys. Rev.* **D34**, (1986) 969.
18. B. Kayser, N. Deshpande, and J. Gunion, in *Neutrino Mass and Low Energy Weak Interactions*, edited by V. Barger and D. Cline (World Scientific, Singapore, 1985), p. 221.
19. B. Kayser, *Phys. Rev.* **D30**, (1984) 1023. B. K. thanks A. S. Goldhaber for the conversation in which this argument was constructed.
20. It has recently been shown that the magnetic moment of a Dirac neutrino tends to be proportional to its mass (hence small) even in the presence of right-handed currents. See J. Liu, *Phys. Rev.* **D35**, (1987) 3447.
21. R. Shrock, *Nucl. Phys.* **B206**, (1982) 359.
22. It has been suggested that if the ν_e has a magnetic dipole moment *much* larger than that expected in the standard model or in popular alternatives, its interaction with solar magnetic fields could reverse its spin, turning it into a noninteracting right-handed particle. This could explain the apparent depletion of ν_e flux from the sun. See M. Voloshin and M. Vysotskii, *Sov. J. Nucl. Phys.* **44**, (1986) 544; L. Okun, M. Voloshin, and M. Vysotskii, *Sov. J. Nucl. Phys.***44**, (1986) 440, and *Sov. Phys.* JETP **64**, (1986) 446.
23. B. Kayser and A. S. Goldhaber, *Phys. Rev.* **D28**, (1983) 2341.
24. J. Nieves, *Phys. Rev.* **D26**, (1982) 3152.
25. B. McKellar, Los Alamos Report N^0. LA-UR-82-1197 (unpublished).
26. E. Radescu, *Phys. Rev.* **D32**, (1985) 1266.
27. J. J. Sakurai, *Advanced Quantum Mechanics* (Addison-Wesley, Reading, Massachusetts, 1967).
28. Our treatment of charge congugation in the Dirac equation follows that of Ref. 27.

29. B. Kayser and R. Shrock, *Phys. Lett.* **112B**, (1982) 137.
30. Special Feynman rules for processes involving Majorana particles have been given in H. Haber and G. Kane, *Phys. Reports* **117**, (1985) 75, and in E. Gates and K. Kowalski, *Phys. Rev.* **D37**, (1988) 938.
31. The dependence of this amplitude on the CP parities of the neutrinos has been discussed in L. Wolfenstein, *Phys. Lett.* **107B**, (1981) 77, and in Ref. 23. See also B. Kayser, Ref. 19, and S. Bilenky, N. Nedelcheva, and S. Petcov, *Nucl. Phys.* **B247**, (1984) 61.
32. D. Chang and P. Pal, *Phys. Rev.* **D26**, (1982) 3113.
33. D. Caldwell *et al.*, *Phys. Rev. Lett.* **59**, (1987) 419. In extracting the bound on M_{eff} from that on the double β decay rate, these authors have used nuclear matrix elements presented in W. Haxton and G. Stephenson, Jr., *Prog. Part. Nucl. Phys.* **12**, (1984) 409, T. Tomada, A. Faessler, K. Schmidt, and F. Grümmer, *Nucl. Phys.* **A452**, (1986) 591, and K. Grotz and H. Klapdor, *Phys. Lett.* **153B**, (1985) 1.
34. P. Langacker, B. Sathiapalan, and G. Steigman, *Nucl. Phys.* **B266**, (1986) 669.
35. M. Fritschi *et al.*, Ref. 6.
36. B. Kayser, S. Petcov, and S. P. Rosen, in preparation; B. Kayser, in *New and Exotic Phenomena*, edited by O. Fackler and J. Trân Thanh Vân (Editions Frontieres, Gif-sur-Yvette, France, 1987), p. 349; B. Kayser, in *Proceedings of the XXIII International Conference on High Energy Physics*, edited by S. Loken (World Scientific, Singapore, 1987), p. 945; S. Petcov, in *'86 Massive Neutrinos in Astrophysics and in Particle Physics*, edited by O. Fackler and J. Trân Thanh Vân (Editions Frontieres, Gif-sur-Yvette, France, 1986), p. 187; B. Kayser, in *Proceedings of the Oregon Meeting (Annual Meeting of the Division of Particles and Fields of the American Physical Society)*, edited by R. Hwa (World Scientific, Singapore, 1986), p. 397.
37. That any such contribution vanishes with the neutrino masses due to an orthogonality relation of the form (4.16) was noted earlier in the case of left-right symmetric gauge theories by T. Kotani, in *Proceedings of the 1984 Moriond Workshop on Massive Neutrinos in Astrophysics and in Particle Physics*, edited by J. Trân Thanh Vân (Editions Frontieres, Gif-sur-Yvette, France, 1984), p. 397. That such orthogonality relations should be more general has been stated in M. Doi, T. Kotani, and E. Takasugi, *Prog. of Theo. Phys. Supplement* **83**, 1985. The argument we have just gone through shows that indeed these relations *are* general.
38. J. Schechter and J. Valle, *Phys. Rev.* **D25**, (1982) 2951; E. Takasugi, *Phys. Lett.* **149B**, (1984) 372.

39. D. Caldwell *et al.*, Ref. 33.
40. For a more accurate estimate of this limit, see R. Mohapatra, *Phys. Rev. D* **34**, (1986) 909.
41. J. Pati and A. Salam, *Phys. Rev.* **D10**, (1974) 275; R. Mohapatra and J. Pati, *Phys. Rev.* **D11**, (1975) 566, 2558; G. Senjanovic and R. Mohapatra, *Phys. Rev.* **D12**, (1975) 1502; R. Mohapatra and G. Senjanovic, *Phys. Rev. Lett.* **44**, (1980) 912, and *Phys. Rev.* **D23**, (1981) 165.
42. We thank G. Senjanovic for teaching us to follow this procedure.
43. G. Beall, M. Bander, and A. Soni, *Phys. Rev. Lett.* **48**, (1982) 848.
44. J. Bagger, S. Dimopoulos, E. Masso, and M. Reno. *Nucl. Phys.* **B258**, (1985) 565.
45. J. Rosner, *Comm. Nucl. Part. Phys.* **15**, (1986) 195.
46. For additional discussion of the general theory of massive neutrinos, see S. Bilenky and S. Petcov, *Rev. Mod. Phys.* **59**, (1987) 671.
47. This last insight is due to S. Nussinov, *Phys. Lett.* **63B**, (1976) 201.
48. This constraint, and its role in determining the vacuum mixing angles and neutrino masses which can explain the solar neutrino puzzle, are discussed in W. Haxton, *Phys. Rev. Lett.* **57**, (1986) 1271.
49. The exact value of Γ_M/Γ_D for all β may be found in S. Rhie, in *Proceedings of the International Symposium for the Fourth Family of Quarks and Leptons*, edited by D. Cline and A. Soni (New York Academy of Sciences, New York, 1987).

ACKNOWLEDGEMENTS

One of us (B.K.) would like to enthusiastically thank R. Cahn, P. Carruthers, A. S. Goldhaber, W. Haxton, S. Petcov, S. P. Rosen, G. Senjanovic, R. Shrock, L. Stodolsky, and L. Wolfenstein for very helpful discussions of neutrino mass over the years. He would also like to warmly thank R. Turlay for creating the Neutrino Mass Study Week at Saclay, and for inviting him to lecture during that week. Finally, he would like to acknowledge the excellent hospitality of the Institute of High Energy Physics in Beijing and the Aspen Center for Physics, where parts of these lecture notes were completed.